# 虚实之间

## ——2021时装艺术国际展·中国西樵

中国纺织工程学会　吕越　主编

中国纺织出版社有限公司

# 内 容 提 要

在百年未有之大变局的当下，新冠疫情在世界范围内产生了广泛而深远的影响，后疫情时代，在生命与自然之间、现实生活与虚拟科技之间，人类如何更好地栖居于我们赖以生存的地球？面对未来超越现实的世界，艺术家如何以时装艺术创作对此做出回应？本画册基于时装艺术家以"生命""自然""科技"为主题，以面向当下的现实与超越现实的未来虚拟世界的创作为主旨，融入西樵地方特色元素，面向全球艺术家征集遴选，将此次"虚实之间——2021时装艺术国际展中国西樵"的展览作品以画册出版的呈现。既是对2007年举办至今的时装艺术国际展及每届展览画册出版的延续，更是对面向当下现实世界的人的关照与对未来超越现实的、与虚拟共生世界的无限期待！

**图书在版编目（CIP）数据**

虚实之间：2021 时装艺术国际展中国西樵 / 中国纺织工程学会，吕越主编 . -- 北京：中国纺织出版社有限公司，2021.12

　　ISBN 978-7-5180-9191-1

　　Ⅰ. ①虚… Ⅱ. ①中… ②吕… Ⅲ. ①服装设计—作品集—世界—现代 Ⅳ. ① TS941.28

　　中国版本图书馆 CIP 数据核字（2021）第 254085 号

责任编辑：宗　静　　特约编辑：李　娟
责任校对：王蕙莹　　责任印刷：王艳丽

中国纺织出版社有限公司出版发行
地址：北京市朝阳区百子湾东里 A407 号楼　邮政编码：100124
销售电话：010—67004422　传真：010—87155801
http://www.c-textilep.com
中国纺织出版社天猫旗舰店
官方微博 http://weibo.com/2119887771
北京华联印刷有限公司印刷　各地新华书店经销
2021 年 12 月第 1 版第 1 次印刷
开本：787×1092　1/16　印张：18.5
字数：463 千字　定价：268.00 元

凡购本书，如有缺页、倒页、脱页，由本社图书营销中心调换

# 目　录

# Contents

# 致辞与评论

## 超越现实的现实
### 琴基淑

时装艺术是将时装的艺术特征发挥到极致的造型创作。通过时装艺术国际同盟的时装艺术展活动，我们一直在发布我们的时装艺术作品，通过这项活动我们彼此分享了很多灵感和情感，并相互学习、启迪和鼓励。事实上，通过与海内外艺术家的会面，我们在时尚与艺术、友谊与生活等方面进行了更多的交流与联系。通过时装艺术，我们也发展和成熟了自己。

然而，2019年开始的全球疫情给时装艺术活动带来了巨大的危机。海内外艺术家奔赴各地活动、举办展览是不可能的，必须探求新的方法。2020年的定期展通过互联网举行。通过用视频影像的形式来代替受到现实制约的现场展览，我开始思考各种方法，为此也有过担心和苦恼。

尽管如此，我们还是庆幸现代科技提供的数字环境正在发挥作用。21世纪，已经是生活中不可或缺的互联网环境让地球更近了，成为我们与海内外朋友建立交流和联系的渠道。超越人类物理极限的时空间契机纷至沓来，超越物理空间的边界，扩展到了虚拟的现实，进入了一个可以实现现实生活的虚拟空间。人们认为，科学技术的发展需要大量时间才能在整个国际社会上传播和共享，但矛盾的是，Covid-19疫情的流行却加速了科学技术的传播和解决问题的共享。

已经全球一体化的世界各国，不再直接的相互往返，面对面接触，而是加速数字环境来解决他们面临的问题。与现实相似的虚拟现实和虚拟空间被更加积极地激活应用。仅限于特定地点的人与人之间的联系在虚拟空间中也得以实现。我们在所处的环境中可以与世界上的任何人建立联系。人类可以在超越物理界限之外的现实中活跃，无处不在。

在体验无处不在的领域的同时，我们也面临着另一个"超越现实的现实"，这次展览将是一个重要的机会，让我们思考疫情之后要做什么，要怎样去做。专注于时装艺术，也可以让我们超越现实的境界，时尚艺术能否给文化、人类、环境带来一些价值观上的改变。因此，在筹备本次展览的过程中，我们相信海德格尔说的"在世存在"（ being in the world ），希望以时装艺术的艺术性，净化自然，治愈人类。我要感谢时装艺术国际同盟的主席吕越和组织这次展览的筹备委员会，也感谢中国纺织工程学会、广东省佛山市西樵镇人民政府、中华服饰文化研究会和中央美术学院时装艺术研究中心的支持使之成为了可能。我为每个人的平安祈祷。

2021年11月
琴基淑　教授
时装艺术国际同盟顾问
韩国弘益大学教授
韩国柳琴瓦当博物馆副主任
曾任韩国服装协会顾问、韩国时尚与文化协会顾问

# Reality beyond Reality
## KeySook Geum

Fashion art can be defined as a work of plastic art that emphasizes the artistic characteristics of fashion, which is an object placed on the human body. Through the activities of the International Fashion Art Network(IFAN), artists have presented their fashion art works, and have shared a lot of learning, inspiration, encouragement and emotion with each other. In fact, by encountering artists from home and abroad, we, the artists have communicated about the fashion art and have connected with friendship, as well. Throughout the fashion art activities, the artists have developed and matured.

However, the global pandemic that began at the end of 2019 brought a great crisis to the fashion art field as well. The IFAN exhibition in which artists and works directly participate could not be held. As a result of exploring various alternatives, the 2020 regular exhibition was held online.Nevertheless, we are grateful that the digital environment provided by contemporary  science and technology is working. This is because the Internet environment, which has been built since the 20th century, has brought the earth closer and has become a channel through which to communicate and connect with the global friends. Now, we have been provided with a platform of another dimension that can transcend the physical limits of human beings, such as place and time.

Reality transcended physical space and expanded into virtual space, and the virtual reality came to be recognized as another dimension of reality. It was expected that it would take a lot of time for this level of advanced science and technology to be spread and shared throughout the international community. Paradoxically, the Covid-19 pandemic has contributed to accelerating the spread and settlement of the new platform.

Many countries in the globalized world have tried to speed up building a digital environment instead of going back and forth to solve the problems they are facing. Virtual reality and virtual space, which are similar to reality, were more actively activated. The connection between humans which was limited to a specific place was realized in a virtual space. We are in an environment where we can connect with anyone in the world. Humans have come to experience the ubiquity of being active in another reality that transcends reality.

Through the enlightenment of omnipresence, which is believed as the realm of God, we realize another reality that transcends reality, provides us with a new perspective. Now, as we prepare this exhibition, we will take it as an important opportunity to think about the direction and role of the Fashion Art after the Pandemic. Even if another reality beyond reality is given to us, the values of nature, people, culture and Fashion Art will not be changed. While the exhibition was preparing, we, as one of the being-in-the-world, hope and expect to purify nature and heal humans with the artistry of fashion art.

It is pleased to express my gratitude to the organizations and institutions such as China Textile Engineering Society, Xiqiao Town People's Government of Foshan City, Guangdong Province, The Society Of Chinese Historical Costume, Fashion Art Research Center of Central Academy of Fine Arts for their supports. I am also pleased to be able to thank the president of IFAN Lyu Yue (Aluna) and committee members for their hard work in planning and running this exhibition and possible. Above all, congratulations go to the participating artists who made this exhibition shine. I wish you all good luck.

November, 2021
KeySook Geum, Ph.D.
Advisor,International Fashion Art Network
Professor, Dept.of Textile Art Fashion Design, Hongik University
Co-director of YOOGEUM Museum of Roof-End Tile
The Former Adviser, the Korea Society of Costume
The Former Adviser,the Korea Fashion &Culture Association

# 为"虚实之间"增彩
## 许平

由琴基淑、吕越等一批艺术家发起的"时装艺术国际展"已经做了十多年了。在这短短的十多年间，"时装艺术"，从一个鲜为人知的陌生概念，变身为一股吸引众多关注、认同甚至资源投入的艺术新潮，这颗在东方的大地上成长并绽放的种子，终于有了哪怕是短暂的属于自己的在线空间，可喜可贺。艺术家们曾经为此而长期地努力，即使在疫情反复的2021年，这种努力也未停止。在"树欲静而风不止"的"抗疫"年代，只要疫情的阴影尚未最终退出社会健康治理的核心关注，在线或部分在线就是现实有效的自由集结方式、参展艺术家可以在一个更富活力的空间里，展示团结与自信。如同风雨之后的芳草地，那种绿意的呈现，更加令人感佩和心动。

今年的展事以"虚实之间"为名，这是个很好的选题，虽然要做好并不易。时装艺术之于国计民生的价值本身就在"虚实之间"。她在大起大伏、汹涌澎湃的纺织工业、服装产业、时尚市场大潮面前，如同千军万马之中的轻骑、交响乐章之中的小号，难有黄钟大吕那般庄严，却可以有清扬明沏的穿透。"实"则可以丰富体验、引领生活装扮的明天；"虚"则可以激发想象、登攀精神创造的殿堂。每届"时装艺术国际展"的幕起幕落，都是这种"虚实"艺术空间张合的盛大仪式，也是这种"虚实"价值客观与否的自我拷问。

回顾十多年的时装艺术国际展，每年数以百计的参展作品，都会有一个不约而同地特征，那就是不同程度的形式象征性；每件作品的背后，都有一个自在的象征主题的生成。这种主题生成的"真值"性，才是令我感到时装艺术之所以作为"时装艺术"灵魂的意义所在、价值所在、精神所在。尽管迄今为止，时装艺术对此尚没有一个明确的定义，但可以从参展的经典作品中得到某种映证。对此我仍想以琴基淑教授在"时装艺术国际展"举办初期的参展作品"舞衣"为例。

这件作品已问世多年，其后琴基淑教授也不断有作为同一系列的衍生新作推出。但对其源头，甚至可称"时装艺术国际展"样式源头的代表作"舞衣"，我仍然认为有继续加以关注的必要，因为对这件作品的美学品质与历史贡献，我们还缺乏必要的理论总结，而这种关注和总结对于发展到今天的时装艺术恰好是十分必要的。

今天看来，"国际展"举办初期展出的"舞衣"作品，堪称完美地诠释了时装艺术的"精神开放性"和"形式开放性"的双重特征。精神开放性是指作品所蕴含的精神象征性，是每位观者立于其前都可以以各自不同的方式接受、感触到的那种奇妙的意蕴传达性。而形式的开放性则产生于作品所选择的有着强烈材质暗示却又产生截然不同反观效果的"舞衣"编织架构，半开合型的"舞衣"形式从空中悬挂于壁，似衣而非衣，有形而无形，仿佛从无际的空气中抽象、集聚、凸显而来，又仿佛向大千世界的深处消融、隐身、抽空而去。这种衣装形式"在"与"不在"的矛盾形式，是对千百年来人与衣着形式的"一体"关系与"异体"关系的深刻反问，也将这一物质形式的审美内涵提到一个新的高度。这件作品不仅从形式上准确把握了时装艺术亦虚亦实、亦真亦幻的特征，而且直观地、富于启示地完成了一个精神象征与形式象征双重叠合的内涵架构，展示了"时装艺术"何以"以小搏大"、何以在有限的物质形式中连接无限的精神天地的可能。相比较而言，"舞衣"作品没有刻意强调浩大的铺排、豪华的场景，甚至也没有刻意强调手工的精致、科技的炫酷，而只是在一种近于严苛的单纯中，将材料异构的趣味魅力与作为"时装艺术"最基本的创意源头与基础平台的"人形"表现力发挥到最大。这些看似具体、个别的表现方法，其中或许蕴含着可以从中抽取出来的某些艺术创作原则。

我个人认为，"舞衣"作品某种程度上代表了"时装艺术"概念提出以来的数十年间在东方所达到的最高艺术成就，同时其艺术语言的独特性、其艺术内涵的无可替代性，已经印证并正在印证着这一艺术类型有可能独立于世界艺术之林的表现性特征。"舞衣"作品对时装艺术内涵的准确把握及其物质形式的绝对契合，其表达手法的自洽性、纯粹性，就视觉表达效果的完美性、独特性，在这一领域极具代表性。十多年来不断成长和扩大的时装艺术圈内，涌现了诸多可圈可点的优秀作品甚至代表作，将来应该逐件地予以关注并进行理论的总结，从而形成作为领域的发展所必需的理论支撑。而"舞衣"在其中代表着某种核心的精神指标性。

虽然"舞衣"已是旧作，拿到今天是旧话重提。如前所述，这不仅是因为本届展览"虚实之间"的主题恰好切中"舞衣"作品以"时装艺术"之名所承载的种种内涵，还因为这十多年来，这件作品几乎已经内化为我内心识别、评判、赏鉴"时装艺术"的一把标尺。"时装艺术"作为一个独立的艺术类型，需要在艺术之林中立身、站稳和形成内涵高度，就必须确立类型本体的艺术标准和品质代表性。"舞衣"作品在这一艺术领域所达到的品质上的无可替代性，是这一艺术类型业已成熟的标志，也是这一艺术领域应有的骄傲。"舞衣"作品的精神象征性表达所具备的超越性价值，十多年间无出其右。其独到性的秘诀在于其纯粹性。换言之，其精神内涵的表达是独立的、完整的，不附加任何其他的标签属性，去掉了一切地域的、民族的、历史的、产业的、观光的、文创的等标签符号，而使作品真正成为人类性的。虽然这些符号也很重要，附加的价值都可以在其他衍生形式中涵括再现，但在"这一件"作为类型"真值"源头的作品形式中，它只集中于体现"人与穿着形式的一体性与异体性、服装作为一种独立的人格文化的自在性与反喻性"这样的主题特征；这种令观者会不由自主地产生"被击中"与"被穿透"的震撼力，不管是今天的时装艺术，还是未来的设计艺术都同样的至关重要。这是来自艺术形式、艺术语言本身的传播价值的种子，艺术与其他需要相结合的社会功能都只不过是这种价值基础的再创造与再加工，前提是内涵种子的真实存在。以现在乳制品市场的"空包"产品为例。"空包"是指以工业化手段催生的、不具有真正奶牛母乳价值只是形态相似的牛乳产品。今天的时装艺术需要杜绝这样的"空包"化成果，努力创造具备时装艺术"真值"的时代作品、原型作品，为立身于"虚实之间"的时装艺术增值、增彩、传神。

2021年11月

许平　教授

国务院学位委员会第六届（艺术学）学科评议组委员、第七届（设计学）学科评议组召集人

北京设计学会会长

中央美术学院教授、博士生导师

# Adding some different perspectives to the International Fashion Art Exhibition "Reality Beyond Reality"
## Xu Ping

The International Fashion Art Exhibition initiated by a group of artists such as Prof Aluna Lyu, Dr KeySook Geumhas been going on for more than ten years. In this short period of more than ten years,, "fashion art" has transformed from a little-known and unfamiliar concept to a new artistic wave that has attracted wide attention, recognition and even investment. This seed that grows and blooms on the Eastern land finally has its own online space even for a short time. Congratulations! Artists have made long-term efforts to do this and even do not stop in 2021 with recurrent epidemic. As the old Chinese saying that "the tree may crave calm, but the wind will not drop", as long as the shadow of the epidemic has not finally withdrawn from the core concern of social health governance, online or partially online is an realistic and effective way of freedom gathering under the anti-epidemic era. The participating artists can show their unity and self-confidence in a more dynamic space.Just like the grassland after a gust of wind and rain,the greeneryis even more admirable and exciting.

The theme of this year's exhibition is "Reality Beyond Reality", which is a good topic, although it is not easy to do well. The value of fashion art to the national economy and the people's livelihood lies between "reality" and "beyond reality". With the rapid developments in the textile and garment industry and fashion market, fashion art is like a trumpet in symphony with instinctive penetration power in which it cannot possess the solemnity like "Huangzhong Dalv" (an Chinese idiom describes music or speech solemn, upright, or magnificent.) but has strong power penetrating the heart. Its "Reality" can enrich the life experience and lead the direction of the future, while "Beyond Reality" can inspire imagination and step into the spiritual palace of art creation. Every "International Fashion Art Exhibition " is a grand ceremony for the art space as well as a self-torture of whether the value of "Reality and Beyond Reality" is objective or not.

Looking back on the International Fashion Art Exhibition in the past years, hundreds of exhibited works were of the same character, that is, they all have different degrees of symbolism: there is a free symbolic theme behind each piece of work. The "true value" generated by the symbolic theme makes me feel the significance, value and spirit of fashion art as the soul of "fashion art". Although there is no clear definition of fashion art so far, some evidence can be obtained from the classic works in this International Fashion Art Exhibition. In this regard, I would like to take Dr KeySook Geum's work "Dancing Clothes" which was exhibited in the early stage of the International Fashion Art Exhibition as an example.

This work has been published for many years, and since then Dr. KeySook Geum has been continuously presenting a series of derivative works with the same theme in the following years. However, I still think it is necessary to keep our focus on their origin, this representative masterpiece "Dancing Clothes" that can even be regarded as the origin of the International Fashion Art Exhibition. The aesthetics quality and the historical contribution of this work have not been summarized from a theoretical perspective, andthis kind of attention and summary plays a crucial role in today's development of fashion art.

In the current perspective, the "Dancing Clothes" exhibited at the initial stage of International Fashion Art Exhibition perfectly interpreted the dual characteristics of underlying fashion art, which are "spirit openness" and "form openness". Spirit openness refers to the spiritual symbolism contained in the work, which can be received and felt in different ways by viewers. The openness of the form comes from the woven structure of the "Dancing Clothes" that has strong materials hints but presents an entirely different effect. The half opening "Dancing Clothes" was hung along the wall, and it brought the thoughts of whether it was a piece of clothes, or it is intangible. The work seemes to abstract, gather, and stand out from the boundless air, and to dissolve, hide and evacuate to the depths of the boundless world. The contradictory clothing form of the "being" and "not being" is a profound rhetorical question of the "unity" and "heterogeneity" relationship between people and the clothing form for thousands of years, and raises the aesthetic connotation of the material to a new level. This work not only accurately grasps the characteristics of the real and fiction in the perspective of the form, but also intuitively inspires the structure of combining the spiritual and formal symbols, showing how the fashion art can deliver a grand idea with small things, and challenge the possibility of connecting the infinity with limited materials and forms. Comparatively, "Dancing Clothes" does not deliberately emphasize a great arrangement, a luxury scene,the fineness of craftsmanship, or the coolness of technology. But in an almost harsh simplicity, it maximizes the interesting charm of heterogeneous materials, the most basic creative source of "fashion art" and the expressive power of mannequin of the basic platform. These seemingly specific and individual expressions may contain some artistic creation principles that can be extracted from them.

In my view, "Dancing Clothes" to some extentrepresents the highest Eastern artistic achievement since the concept of "fashion art" has been raised decades ago. At the same time, the uniqueness of its artistic language and irreplaceable nature of its artistic connotation have reflected this type of art may be independent among the art of the world. "Dancing Clothes" is very representative in this field for its accurate grasp of the connotation of fashion art, its absolute fit of material form, its self-consistency and pure expression methods, and the perfection and  uniqueness of visual expression. The field of fashion art has been growing and expanding continuously for more than a decade. There are many outstanding works and even masterpieces that have emerged. We should pay attention to them one by one and make a theoretical summary, so as to form the theoretical support necessary for the development of this field in the future. "Dancing Clothes" represents some core spiritual indicators in the field.

Although "Dancing Clothes" is a previous work, it's an old talk for today. As mentioned above, this is not only because the theme of "Reality Beyond Reality" is exactly in line with the connotations of fashion art carried by "Dancing Clothes", but also because this work has nearly become a standard for me to recognize, judge and appreciate fashion art in the past ten years. As an independent art form, "fashion art" needs to establish its artistic standard and quality representation like many other art types and forms a high level of connotation. The irreplaceable quality of "Dancing Clothes" is a sign of the maturity and the pride of fashion art. The transcendence value of the spiritual symbolic expression of "Dancing Clothes" has been unsurpassed for more than a decade. The secret of its uniqueness lies in its purity. In other words, the expression of its spiritual connotation is independent and completed, without attaching any other label and removing all labels that apply to regional, ethnic, historical, industrial, sightseeing, cultural, creative, etc. that makes the work human empathy. Although these symbols are very important and the added value can be formed and reproduced inother derivatives, in "this true value" form, it only focuses on reflecting in "the unity and heterogeneity of human and way of dressing, and the liberty or satire when clothing exists as an independent culture". This shock that makes the audience having the feeling of "being hit" or "being penetrated" involuntarily is equally important for current fashion art and future design. This is the seed of communication value from the art form and art language itself. The social function of combining art with other needs is nothing more than the re-creation and re-processing of this established value, provided that the real existence of the connotation seeds. For example, the "empty pack" is a kind of dairy product selling in the market. "Empty pack" is produced by industrial methods that make it similar to milk form but not real milk. Fashion artists need to end this "empty pack" situation, and strive to create works and prototypes with the "true value", to make the fashion art be with rich meanings and dynamics then beyond reality.

Xu Ping , Ph.D.
Member of the 6th municipal discipline appraisal (Arts) group and Convenor of the 7th municipal discipline appraisal (Design) group, , Academic Degrees Committee of the State Council
Chairman, Beijing Design Society
Professor and doctoral supervisor ,Central Academy of Fine Arts

# 时装艺术的又一次绽放
## 吕越

"时装艺术"（Fashion Art）作为艺术的一个分支，虽然与其他当代的艺术形式同时诞生，但她却在时尚界大放异彩，不仅影响着时尚T台，也引领着时尚思潮的流变。"时装艺术"因其具有艺术的敏感特性，长期在艺术的前沿思维、媒介创新等方向深耕，在面对人类世界出现的新问题时，迅速以作品来进行回应。此次的展览主题既顺应时代发展，又给予了时装艺术家们在应对现实问题的发声机会。

今天的时装和时尚市场已经呈现出对于更新时间上愈加迅速的创新诉求，设计师仿佛一开始便被赋予了快速出新的"特殊使命"。"求新求变"从来都是时装设计的基因，是其最具"折腾"的特性，它包含了对某种社会预期心理的领先创造，它要求设计师感受"新"的触角必须变得敏感。时装设计也开始借助艺术的力量尝试提供创新的样本，示范设计如何破除边界的壁垒，试探设计还能怎样求新求变。于是"时装艺术"带着特殊的背景和身份走进社会生活，以更为大胆的视角和方式为设计打开尝鲜的大门，它面对现实，也塑造理想。

面对后疫情时代，我们正在经历着世界格局的全新变化，也见证着生存环境的调整演变，艺术家回应着如何治愈伤痛，也思索着如何诠释未来。伴随后疫情时代，"虚拟时代"的到来在这场变革中显得尤为迅捷而又广泛。数字化的交流与交易给了民众更多元的参与到现代时尚生活中的可能性，"非实体化"的生活方式正在影响着时装艺术和时尚传播的蜕变方向。时装艺术从诞生之时起，便为的是突破边界、寻求引领，这也是时装艺术家们在数次回应时代诉求时所作出的努力。

"时装艺术国际展"在中国自2007年发展至今，已成功举办十余届国际性的大型展览及学术研讨、出版等系列学术活动。而随着"时装艺术"的实践和理论逐渐趋于成熟，汇聚了国内外众多热爱这一艺术表现形式的创作者，2015年"时装艺术国际同盟"（International Fashion Art Network）应运而生。时至今日，它们已经伴随中国民众走过了十余年的时尚历程。在经历了受疫情影响的2020年线上展览之后，"2021时装艺术国际展 中国西樵"带我们重回线下，艺术家们将时装艺术结合西樵地域文化展开主题创作，开启全新的西樵文化之旅。

本次有120位艺术家作品入选，来自于中国、英国、法国、德国、匈牙利、墨西哥、加拿大、美国、巴基斯坦、韩国等十余个国家和地区的不同文化背景的艺术家贡献了二百余件作品。他们以"虚实之间"为创作主旨，共同针对"人"在当下现实与未来虚拟之间，面向"生命、自然、科技"的发问并与之进行探讨。

此次展览，将以线下和线上两种方式同时呈现，借用后疫情时代网络发展优势，将现实与虚拟世界链接。并且面向全新主题在创作表现形式上增添了科技艺术与NFT艺术的新方向，是对当下现实美好生活的关切，也是对不断超越现实的未来给予无限好奇与期待！

2021年11月

吕越　教授

时装艺术国际同盟主席

中国纺织工程学会时装艺术专业委员会主任

中央美术学院教授、博士生导师

# Another Blooming of Fashion Art
## Lyu Yue (Aluna)

Fashion art, a branch of art, was although born along with other contemporary art forms, blooms in the fashion industry. It not only influences the fashion runway,but also leads the evolution of fashion thoughts. Fashion art has sensitive characteristics and has been deeply involved in the cutting-edge of thinking and media innovation. When new problems occurr, artists may express their attitudes to the problems quickly through fashion art. The theme of the exhibition this year conforms to the development of times and also gives fashion artists an opportunity in dealing with practical problems.

Today's fashion industry has shown an increasingly rapid demand for innovation and it seems that fashion designers have been given a "special mission" to create new ones quickly. Fashion design has the gene of "innovation and change", which is its most "dramatic" characteristic. Fashion design leads the creation to meet social expectations and requires designers to become sensitive to the new trend. Fashion designers have also tried to provide creative samples with the power of art, then to demonstrate how to break through the barriers of boundaries as well as explore new possibilities in design. Thus, "fashion art" enters social life with special background and identity and try to open the door of "Innovation" from bolder perspectives and ways. It faces reality and shapes ideals directly.

In the post-epidemic era, we are experiencing brand-new changes in the world pattern and also witnessing the evolution of the living environment. In this sense, artists respond to how to heal the pain but also think about how to interpret the future. Under such a transformation, the advent of the "virtual age" is developing particularly rapid and wide-ranging. Digital communication and transactions have given people more possibilities to participate in modern fashionable life. And the evolution of fashion art and fashion communication is affected by this "non-substantial" lifestyle. Since the birth of fashion art, it has been to break through the boundary and seek its leading position, which is the effort made by fashion artists constantly inresponding to the demands of society.

More than ten related large-scale international exhibitions, a series of academic activities such as seminars, publications have been accomplished since the first International Exhibition of Fashion Art held in China in 2007. With the gradual maturity of the practical experience and theory of fashion art, many artists from home and abroad who are enthusiastic about fashion art have gathered. Then, the International Fashion Art Network was established in 2015 at the right moment. Nowadays, they have accompanied the Chinese people for more than ten years. Affected by COVID-19, an online exhibition was held in 2020. In 2021, International Fashion Art Exhibition comes back to offline, which will be held in Xiqiao, China. The participating artists will combine fashion art with the local culture to create new artworks and bring brand-new cultural developments to Xiqiao.

Total 120 artists from different countries with different cultural backgrounds submitted more than 200 artworks to this exhibition. They are from China, the United Kingdom, France, Germany, Hungary, Mexico, Canada, the United States, Pakistan, South Korea, etc. The artists take the exhibition theme, "Reality Beyond Reality", to explore and discuss the connections among life, nature, and technology, as well as the situation of "human" living between the present and the future.

The exhibition presents online and offline simultaneously. With the advantages of the internet, it links the real and virtual worlds together in the post-epidemic era. Meanwhile, the exhibition embraces new forms of science and technology art and NFT art for creative expression, which is not only a concern for the beautiful current real-life but also delivers curiosity and expectation to the future beyond reality.

Lyu Yue(Aluna), Ph.D.
Chairman, International Fashion Art Network
Director, Fashion Art Professional Committee of China Textile Engineering Society
Professor and doctoral supervisor ,Central Academy of Fine Arts

# 时装艺术作品
# Fashion Art Works

An Da

安达

This series uses the characteristics of knitting technology to express the waves, like a touch of ink brush on the  body. At the same time, the process com bines 3D printing technology, to make a chemical reaction between knitting design and technology.

该系列设计利用针织工艺特性表现海浪波澜，犹如一抹水墨笔画挥洒于人体之上；同时工艺结合3D打印技术，实现针织设计与科技的一次新鲜化学反应。

Sea
Wool, silica gel

波澜不惊
羊毛、硅胶塑料

Andrea Benahmed Djilali (Hungary)

安德莉亚 · 本纳迈德 · 吉拉利（匈牙利）

"When I design and create my soul starts to heal"
Design and creating  a handbag for me are a form of
art where I can get away with expressing myself in
my truest form. It is related to my emotional.
Meditation such as mindfulness, or focusing the mind
on a particular object, colours or techniques–to train
attention and awareness, and achieve a mentally
clear and emotionally calm and stable state. Wearing
a handbag you love can do wonders for your mood
and confidence.
With this in mind, I set out to design the kind of
handbag that has the ability to make my consumer
stand out from the crowd.

"当我设计和创造时，我的灵魂将被治愈。"
设计和创作个手包对我来说是一种艺术形式，我可以逃离
并用最真实的方式表达自己。这与我的情绪有关。
正念或将注意力集中在特定的物体、颜色或技巧上，这类
冥想可以训练注意力和意识，并达到思绪清明、平静、稳
定。背上一个你喜欢的手提包，可以给你带来惊喜和自信。
考虑到这一点，我开始设计一种手包，可以让购买它的人
可以从人群中脱颖而出。

Myrna Marble Art
Vegetable tanned leather, gold metallic cork leather,
gold metal magnetic lock, gold chain 120 cm

默娜的大理石艺术
植物鞣皮革、金属软皮、金属磁扣、120 厘米金链

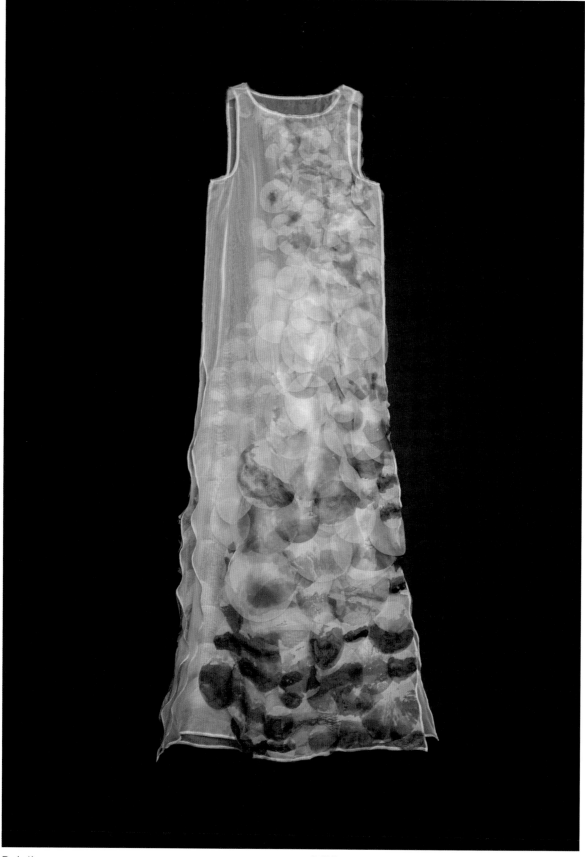

Bai Jingyan

白敬艳

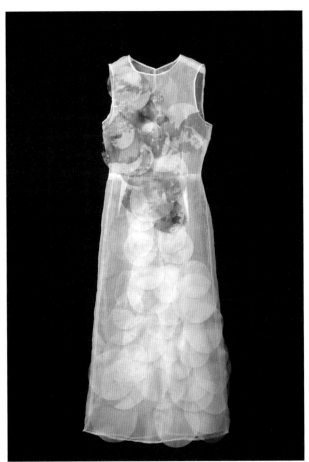
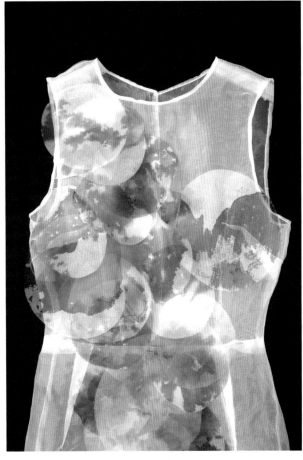

This series mainly uses silk and other yarn materials combined with hand-dyeing to shape the effect of Chinese ink painting. The works take yarn and silk as clothing raw materials. The Silk Organza were cut out into circular pieces and hand-dyed to creat the effect of Chinese ink painting, some circular pieces were selected to make a wet felt treatment together with wool felt, or a drawing treatment. The circular silk pieces are arranged and combined in the form of gradient color, partly hidden and partly visible in the transparent yarn material, match with the tone of Chinese ink painting. This series as a whole show imaginaryeffect, fluidity and feeling of space.

该系列主要是利用真丝等纱质材料结合手工染色塑造水墨画的效果，以真丝欧根纱等为服装原材料，对真丝绡圆片进行手工染色，绘制出水墨画效果，并选取了部分圆片与羊毛毡进行湿毡处理，或进行抽丝。正圆形真丝绡片以渐变色彩形式进行排列组合，通过纱质材料的若隐若现的特性，配合水墨的色调，整体表现出虚幻感、流动感与空间感。

Between Mountains and Rivers
Yarn, silk voile, wool felt, etc.

山水之间
欧根纱、真丝绡、羊毛毡等

"One side soil and water, one side people". Beautiful landscape and a party of good people, together constitute a person's life memory, No matter where they go, the beautiful scenery and the feelings that can not give up. The works hope to use the most simple structure to record conversion when people travel between the mountains and rivers, the emotions, and spaces. Embroidery process, weaving process to express the image of landscape, with different transparency of the fabric expression space on the change. Emphasize People's praise of nature, awe and the desire to blend into nature.

"一方水土，一方人"。美丽的山水和一方纯良的人们，共同构成了一个人一生无法割舍的记忆，无论走遍天涯海角，那份美景，那份情感，都无法割舍。作品希望用最单纯的服装结构来记录人们游走在山水、情感和空间之间的转换。用刺绣工艺、编织工艺表达意象山水，用不同透明度的面料表达空间上的虚实变化。强调人对自然的赞美和敬畏，以及融入自然的愿望。

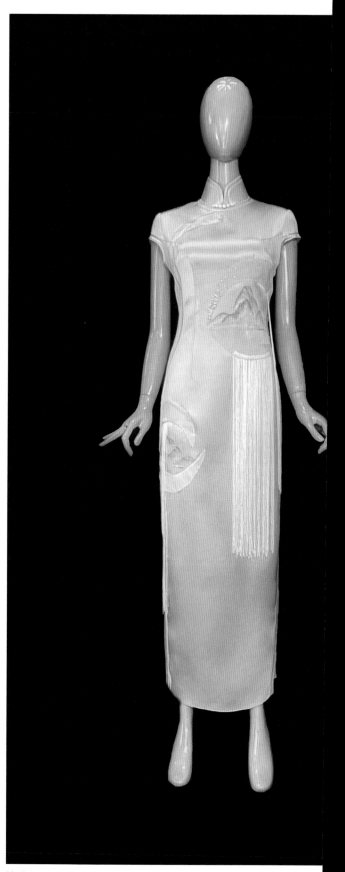

Bao Shuyi

鲍殊易

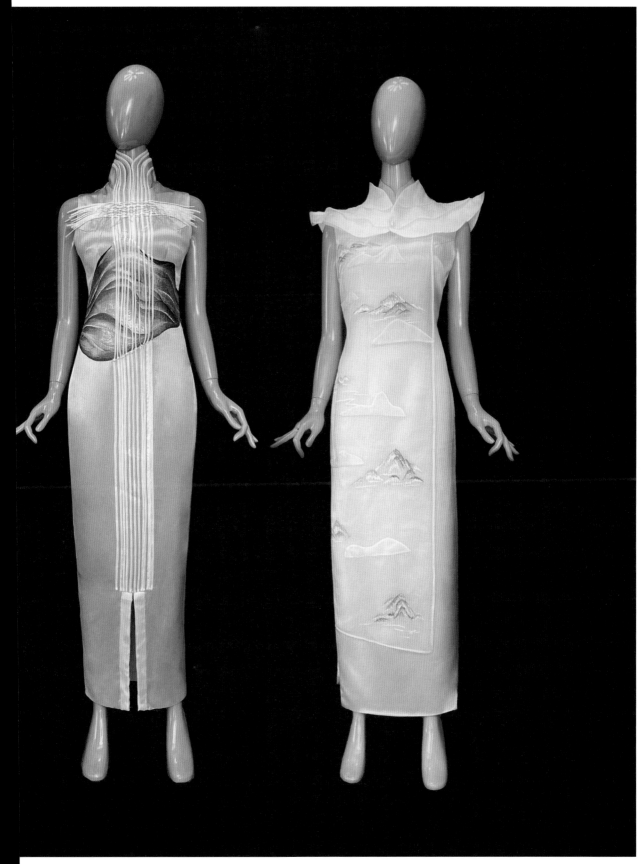

Hometown
Acetate fabric, organza

故乡
醋酸面料、欧根纱

The series use city impression as the theme, focus on Guangzhou, it enables 3D printing, laser cutting and digital printing technology to show many thoughts on urban memory, consumer culture and future changes in the name of fashion and technology.

作品以城市印象为主题，以广州为诠释重点，采用3D打印、激光镭射切割、数码印花技术，以时尚与科技之名，展现出对于城巾记忆、消费文化和未来变革等方面的诸多思考。

Bao Yiwen

鲍怿文

Guangzhou Image
Comprehensive materials

广州印象
综合材料

Carmen Rion (Mexico)

卡门·里翁（墨西哥）

Early in the morning and at night, the light make drawings in my room. Sometimes just lines from the cars that pass by and slip through the curtains, walk across the ceiling and disappear.Sometimes my shadow or my cat's becomes huge, covering the entire wall and even the ceiling, as we grow, we become flat monsters. Shadows come whenever they want, and suddenly disapear too, always depending on the presence of light.

My bed use to be my mother's and is the only thing I asked for in inheritance. Some times my memory as a child feels welcomed by this bed and its white sheets that are no longer the same, nor is the mother receiving on those nights of dark shadows to silence my fears... but the aroma, yes, the memory is recreated and the sensation of the nest, of wrapping myself , feeling contained returns. That is dressing too, dressing in any way, all wraped around with fabrics. There is another life not only in the sheets but in my own hands as I write, others come out that are reflected and accompany me or lash out at me on this keyboard, and that shadow keeps me company while I write. So walking I am accompanied by the other one who walks on the floor and becomes big and small, grows by the walls too and crosses streets encompassing everything and cars do not hurt her, what does the shadow speak of that is mute and only goes with us! on all sides. It's another simultaneous life, another rumor. The day comes and looks trough the window... Shadows rise only when the light shines through. The shadows are co-dependent, they depend on the whims of light, the sun and the moon, moonless nights are dark, all shadows rest, for a moment, how long does the new moon last? "Shadows nothing more between your love and my love." From Javier solis, the king of the ranchero bolero.

清晨和夜晚，光在我的房间里作画。来往车辆的灯光透过窗帘折射进房间，反射在天花板上，又消失了。有时我的身影或我养的猫的影子也会变得巨大，覆盖住整面墙壁甚至天花板，我们仿佛变成了扁平的怪物。影子会随时出现，也会突然消失，这取决于光在哪。

我的床过去是我母亲的，这是我唯一想要从她那里继承的东西。有时候，我童年的记忆会被这张床和它已不再相同的白床单唤起，虽然母亲再也不会像以前那样，在夜里来到我身边，帮我驱赶对黑暗的恐惧……但香气是真切的，记忆被重新创造，如置身巢穴中，将自己包裹，那些被包覆的感觉又回来了。穿衣打扮也是如此，以不拘一格的方式去穿着，用织物包裹身体。当我在写作的时候，影子伴随着我，不仅投射在纸上，也映在手中。这时候其他人也出现在我的脑海中，伴随着我，又或者在我打字的时候在敲打着我。所以走路的时候，我有另一个人陪着，它走在地板上，忽大忽小；也长在墙壁上，穿过包罗万象的街道，汽车穿过也伤害不了它。影子是沉默的，但它与我们共行！这是另一种平行而往的生活，是另一种谎言。白昼来临，透过窗户看去，只有在阳光照射进来时，影子才会出现。它们是相互依存的，影子依赖于光的变换。太阳和月亮，没有月亮的夜晚是黑暗的，所有的影子都会小憩。那新月会持续多久呢？

"影子就在你和我的生活之间"，在你的爱和我的爱之间没有任何阴影——来自哈维尔·索利斯，牧场主的国王。

Recicling Shadows
Linen, silk, cotton

依附于影子
亚麻、丝绸、棉

混沌之光

Cao Yupei / Liu Peilin 曹宇培 / 刘沛霖

The work uses different washed Cowboys for silica gel composite color treatment, and adds recycled plastic materials to represent the environmental problems and increasingly serious pollution. Through the unified comparison and aggravation of colors, the artistic effect is improved by different composite materials. At the same time, strengthen the contrast of order to show the natural treatment of chaos and nature, and form mutual blending effect and mixed multi-dimensional effect.

作品利用不同水洗牛仔进行硅胶复合颜色处理，加入再生塑料材料用来表现日益严重的污染所造成的环境问题，通过色彩的统一对比和逐一加重，用不同的复合材料造成时装的艺术效果提升。同时加强秩序对比来表现混乱和自然的天然处理，在此基础上形成相互交融的肌理效果，和混合的多维效果。

Light of Chaos
Washed denim, silicone, silk, mixed materials of different materials

混沌之光
水洗牛仔、硅胶、真丝、不同材质的混合材料

Chen Ai

陈艾

Formed by the heritage of traditional grass cloth weaving craft materials, combined with modern design elements, the shape of the work is presented with the effect of female dress, can wear but is not suitable for daily wear, expression of the traditional manual weaving has natural property of material texture and links in the form of contemporary fashion, presents the traditional weaving for raw material with ramie woven fabric in the fashion dimension of innovation.

作品由非遗传统夏布织造技艺形成的材料，结合现代造型元素，模特穿着展示作品形态，作品可穿着但不囿于日常着装，表达传统手工织造的具有自然属性的材料质感与当代时装形态的链接，呈现以苎麻为原料的传统编织的平纹布在时尚维度的创新探索。

Primitive & Shape
Grass cloth

原·塑
夏布

Chen Chanjuan (United States)　　　　陈婵娟（美国）

The design was inspired by an image of artic ice broken off from a nearby glacier and melting in the heating ocean. Using a hand bleach approach on the denim fabric and the modular design concept, the designer aimed to highlight both the sublime beauty and the shifting condition of the planet. The method of using 3D prints as modules that connected with denim fabric through 3D printed buttons was a new way of incorporating technology into modular designs as well as increase the durability of the modules.

该设计的灵感来自冰川断裂并在加热的海洋中融化的北极冰的图像。设计师在牛仔布面料上使用漂白方法和模块化服饰设计理念，旨在突出地球的崇高之美和全球变暖的趋势。设计运用了3D打印机打印出来的模块和按钮与牛仔布连接的方法，将传统与现代技术融入模块化的服饰设计并增加模块服饰的耐用性。

Vanishing Ice
Denim and 3D printing

消失的冰山
牛仔面料、3D 打印

Chen Huaxiao                                    陈华小

This series of works are inspired by traditional Chinese cheongsam, with innovative designs. Set off the beauty of harmony between man and nature by using colors with the feeling of transparent, quiet and ethereal. The fabric uses special yarn for secondary fabric reconstruction, which highlights the characteristics of clothing, expresses the integration of virtual and real of the clothing picture, and makes the clothing more layered.

本系列作品的灵感来自中国传统旗袍，加以创新设计，色彩运用通透静谧空灵的感觉，衬托人与自然和谐之美；面料使用特殊的纱质进行二次面料再造，作为服装的特性凸显，表达出服装画面的虚实相融，使服装更具有层次感。

Derivative Shadow
Yarn

衍影
纱类

Chen Wei

陈炜

The series of clothing design is inspired by the Dong culture in Guangxi, combining the plane structure of the Dong traditional clothing with Dong building "Drum Tower", expressing the ethnic embroidery decoration and color aesthetic characteristics in modern clothing language, and focuses on the virtual and the reality by using different clothing materials. This series design not only conforms to the current fashion context, but also presents the aesthetic interest of national culture. The series design aims to re-understand and express the traditional Dong culture and art in modern clothing language. The redesign of contemporary clothing art is a continuation of the heritage of national culture, showing the bright vitality of traditional national culture to the outside world in a brand-new way of expression.

系列服饰设计以广西地区侗族文化为灵感，把侗族传统服饰的平面结构与侗族建筑"鼓楼"的立体造型相结合，以现代服饰语言表现少数民族刺绣装饰与色彩审美特点，服装材料的运用注重虚实结合丰富视觉层次，使系列设计既符合当下时尚语境，又呈现出民族文化审美意趣。系列设计旨在以现代服饰语言重新理解和表达传统侗族的文化艺术，当代服饰艺术的再设计是民族文化传承的续曲，用全新的表达方式向外界展示传统民族文化的璀璨生命力。

"Dong" Listening
Textile

"侗"听
纺织品

Vincent Chen                     陈闻

Red Garden: The design concept of technology, environmental protection and fashion is fully integrated. The passionate red three-dimensional jacquard fabric is combined with the digital printing fabric of landscape oil painting to create the artistic effect of wearing the scenery on the body while protecting the environment.

The birth of Venus: Combining the design concept of perfect clothing art with environmental protection, European oil painting patterns are displayed on denim fabric by laser technology. Combining the stitching and segmentation of the structure, the three-dimensional effect is achieved to avoid the pollution and waste caused by the washing process of denim.

红色花园：将科技、环保与服装时尚充分融合的设计理念，热情奔放的红色立体提花面料与风景油画数码印花面料相结合，在环保的同时，营造将风景穿在身上的艺术效果。

维纳斯的诞生：将服装艺术完美的与环保结合的设计理念，把欧洲油画图案采用镭射工艺展现在牛仔面料上，结合结构上的拼接分割，达到立体艺术的效果，避免牛仔洗水过程中造成的污染和浪费。

Red Garden / The Birth of Venus
Denim fabric, digital printed denim fabric

红色花园 / 维纳斯的诞生
牛仔、数码印花牛仔

Chen Yanlin                    陈燕琳

The work takes the chrysanthemum among the Four Gentlemen among the traditional Chinese flowers as the theme, and draws on the elegant and simple artistic expression of the Cizhou kiln in the Song Dynasty. It has a clear and elegant, beautiful and vulgar temperament of chrysanthemum, and is decorated with dots of exquisite rice beads. The modern dress style demonstrates a contemporary oriental taste of life, and interprets the ethereal and far-reaching aesthetics pursued by future fashion design realm.

作品以中国传统花卉的"四君子"中的菊花为主题，借鉴宋代磁州窑剔花画瓶的典雅质朴艺术表现手法，重塑经典菊花纹饰，并印制在透明而轻盈的天然真丝欧根纱上，表现出清隽而淡雅、骨秀而绝俗的菊花气质，并装饰着星星点点的精致米珠，以现代礼服造型彰显出当代一种东方生活品位，演绎着未来时尚设计所追求的空灵淡然的美学境界。

Black Chrysanthemum
Organza

墨菊系列
欧根纱

Cheng Hao                           成昊

The twelve flower God is a Chinese folk legend. In China, each flower has its own goddess of flowers, and each has its own beautiful story. With the replacement of the season, flowers are also a variety of different face colorful earth. At the beginning of 2020, the new novel coronavirus (2019-nCoV) outbreak, human beings in the face of disaster is a community of common destiny.The virus will not go away so quickly, it needs to see the "Formation"of good habits of human beings, the virus will not continue to wreak havoc, because human love will gather greater strength to keep the virus away, time will tell us everything and prove what's right. The designer uses batik technology to make clothes from the patterns of the twelve flower Gods, advocating that everyone should still bloom like a flower even in the worst circumstances.

十二花神是中国的民间传说，在中国，百花各有其司花之神，也各拥有一段美丽的故事。随着季节时令的替换，百花也以各种不同的容颜缤纷了大地。2020年初"新冠"疫情暴发，灾难面前人类是命运共同体。病毒不会那么快地离去，它需要看到人类好习惯的"养成"；病毒也不会一直肆虐下去，因为人类的爱会凝聚更大的力量让病毒远离，时间会告诉我们一切，时间也将证明什么才是对的。设计师运用蜡染工艺把十二花神的纹样做成衣服，倡导大家即使在恶劣的环境下也依然要像花朵一样绽放。

The Flowers Bloom in Sequence
Cotton and linen

次第花开
棉麻

Cheng Qi

程琦

Researching into the deconstruction of weaving technology, fabric reconstruction and derivate product design of the Grass Linen, provides the major source of the inspiration for modern designer to build various creative designs and makes traditional craft to exist the space more vastly. Applying modern design and fashion design concept to combine materials and traditional crafts to the Grass Linen. Some methods of tie-dye, batik, embroidery, knitting, to do design. The research results of applying the Grass Linen's traditional element to modern contemporary design, serve our lives.

从夏布织造技艺的解构、夏布的再造、衍生产品设计三个环节进行研究，为现代的设计创作提供丰富的创意源泉，试图用设计为传统工艺找到更广阔的生存空间。重点在于运用现代的设计思路和时尚的设计理念，将夏布与其他材质和传统技艺相结合，如扎染、蜡染、刺绣、编结、首饰制作等进行艺术品创作和夏布作品的系列设计，研究传统夏布元素在当代设计中的运用，并服务于生活。

Chinese Grass
Ramie silk

衍．夏语
苎麻丝

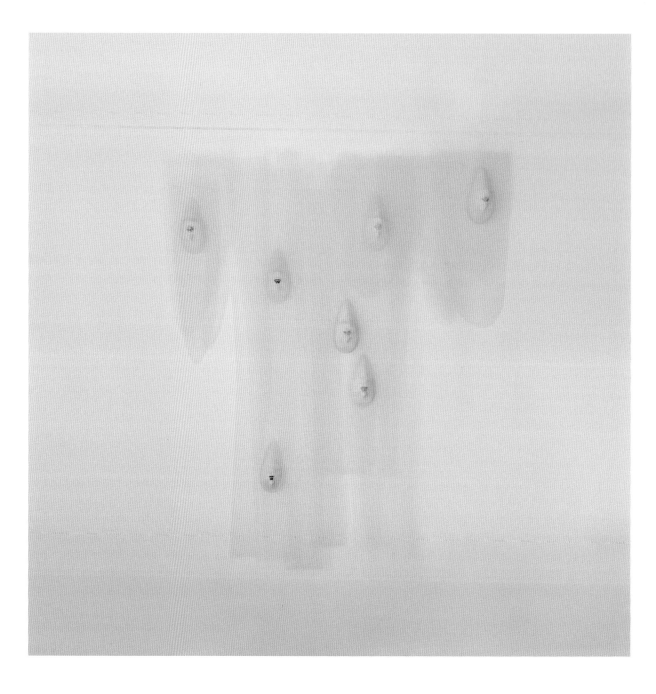

Deng He / Zhou Xuan

邓鹤 / 周璇

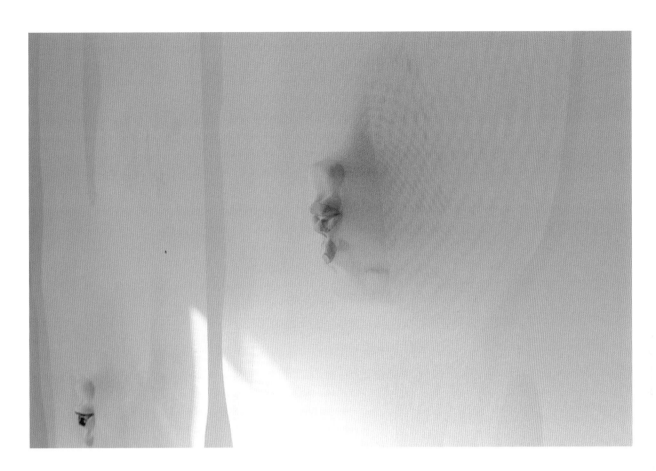

Inspired by Su Hui, the author of *Xuanji picture poetry* in the pre Qin period, it is said that the power of women in the long history of Chinese culture, such as water droplets, looms in the river with layers of curtains.Different from the path from reality to virtual, this work is first realized through virtualization, and then inversely introduced into reality. It rethinks the possibility of real fabrics and forms an observation of the virtual world in real space.

以相传前秦时期《璇玑图诗》的作者才女苏蕙为灵感，诠释中国文化漫漫历史长河中女性的力量，如水滴般的面庞，在层层幔帐的河流中若隐若现。区别于由现实到虚拟的路径，这件作品先通过虚拟化实现，后逆向导入现实，重新思考现实面料的可能性在现实空间中形成对于虚拟世界的观照。

Xuan Ji
Ceramic, gauze

璇玑
陶，网纱

Du Wen

杜文

Using different materials to make the same works, just like the mirror, to form a magical relationship between the similarity and the dissimilarity. I have learned from you and tried to imitate and reproduce with respect and worship, but after all, it comes from different angle of view. Everything breaks between the virtual and the real, with some regrets and some relief. The times are changing unconsciously.

不同的材料制作出相同的作品，就像照镜子，在似与不似之间形成一个神奇的关系，我传习了你，带着恭敬与崇拜，力图模仿再现，但终究来自不同视界，一切都在虚实之间突破疆界，有些许遗憾，也有些许释然。时代在变，在不觉中改变。

Mirror
Wool, organza

镜子
羊毛、欧根纱

Du Yaxi

杜雅茜

The design of *The guide of the exploration in dream* is inspired by the dream during childhood. When she was a child, she always hope herself could grow up quickly, so she often stole the clothes that were designed for adults. The series of clothing use the oversize version to make the sleeves longer design, the fabric reconstruction is in the form of hand-painted rendering design to traditional hand embroidery techniques. The whole series design is focus on integration, with the modern men's wear version design combined with traditional manual embroidery.

《独家梦游指南》是以童年时期的梦境为灵感来源展开设计。小时候总希望自己能够快点长大，所以常常偷穿大人的衣服，系列服装就以Oversize的板型做袖子加长设计，在面料再造方面用手绘的形式绘制图案，再以传统手工刺绣的技法来表现。整体系列设计注重融合，将现代男装成衣的板型设计与传统手工刺绣相结合。

The Guide of the Exploration in Dream Series
Denim, woolen, artificial wool fiber

独家梦游指南　系列作品
牛仔面料、毛呢面料、人造毛纤维

Arvin Gao　　　　　　　　高明

Through reconstruction, color and movement, the design imagines the butterfly dancing gear, endows the fabric design with different power and image, and infuses the original butterfly poetic romance with more charming vitality. All kinds of objects such as immortal flowers of beautiful things and butterflies are mixed with the joyful design of leaping, which gives the elegance and vitality of the fabric design a fresh feeling.

本设计通过重构、色彩和动作，将蝴蝶飞舞具像化，赋予面料不同的力量与形象，给原本的蝴蝶诗意浪漫注入更迷人的生命力。各种各样的物体，如永生花，美丽的物体和蝴蝶以跃动的欢乐设计进行混合，为此套面料设计的优雅和活力赋予新鲜感。

The Butterfly Fly
Silk

蝶·生
丝

Gao Ying (Canada)                    高盈（加拿大）

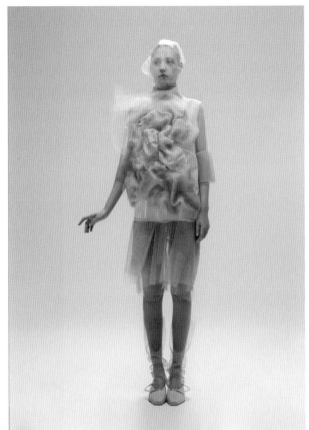

This project was inspired by neurologist Oliver Sacks' novel, The Man who Mistook his Wife for a Hat, in which he relates the story of Jimmie G, a 49-year-old former sailor convinced of being aged 19 since having left the Navy. Shocked by his own reflection when Sacks hands him a mirror, Jimmie reverts to his 19-year-old self as soon as his gaze leaves the reflective surface. Having lost any sense of temporal continuity, Jimmie lives as a prisoner to this single, perpetual moment, oscillating between a presence to the world and a presence to self. Much like Jimmie G, the garments evolve between two states and display perpetual metamorphosis as they react to the presence of the spectator. This travelling between opposite states – from immobility to movement – does not operate as a dichotomy. Upon the field of time, which injects energy into the very core of inertia, fluctuates the intensity animating each garment in its unique way. These two states are mere dropping-off points among an infinite array of possibilities.

Flowing Water, Standing Time
Robotic clothing, silicone, glass, organza, electronic devices

这个项目的灵感来自神经学家奥利弗·萨克斯的小说《误以为妻子是帽子的男人》，他在小说中讲述了吉米·G的故事，他是一位49岁的前水手，自从离开海军后确信自己已经19岁。当萨克斯递给吉米一面镜子时，吉米被镜子里的自己震惊了，当他的目光离开镜子的反射面时，他又回到了19岁的自己。失去了时间上的连续性，吉米像囚徒一样生活在这个单一的、永恒的时刻，在对世界的存在和对自己的存在之间摇摆。这些服装就像吉米·G一样，在两种状态之间进化，随着观众的出现，它们呈现出永久的变形。这种在相反状态之间的旅行——从静止状态到运动状态——这并不是一种二分法。在向惯性核心注入能量的时间场里，以其独特的方式波动着每件衣服的动感强度。这两种状态仅仅是无限可能序列中的落点。

流动的水，静止的时间
机械服、硅胶、玻璃透明硬纱、电子设备

He Ran

赫然

Inspired by the traditional intangible cultural heritage handicrafts of Foshan, the work draws on Lingnan lion awakening culture, lantern technology and unique architectural style. Through the subjective interpretation of Foshan, the "taste" and "memory" of Foshan are expressed in words, and then the words are encoded and encrypted on the fabric. The unique symbols and patterns in culture are reprocessed by means of digital fractal, and the beauty of fashion art and mathematical art is presented by means of repetition, superposition and law. We hope to reinterpret the irreplaceable beauty of traditional technology through modern digital means, and build a virtual and real world through the link between fashion and materials, tradition and contemporary.

作品以佛山的传统非物质文化遗产手工技艺为灵感，汲取了岭南醒狮文化、花灯工艺以及独特的建筑风格等元素。通过对佛山这座城市的主观解读，将佛山的"味道""记忆"等通过文字表述，再将文字进行编码加密的手段呈现在面料之上。文化中独有的符号、纹样等通过数字分形的手段再处理，用重复、叠加、规律等方式，呈现出时装艺术与数字艺术之美。希望通过现代数字化的手段重新演绎传统工艺无可替代的美，通过时尚与材料、传统与当代为纽带，构建一个虚拟与现实的世界。

Memory Coding
Printed fabrics, acrylic, PVC and other comprehensive materials

记忆编码
印花面料、亚克力、PVC 等综合材料

Hu Yuanyuan

胡园园

Dragon, through the difficult robbery can be turned into a real dragon, has a strong power, a symbol of the Chinese strong vitality, and the angel in white, risking his life, retrograde and heroic spirit, Through the epidemic China earth slowly wake up like a sleeping dragon, wake up that moment will shock the world, all the energy in the silence of the savings will usher in the most gorgeous bloom.

蛟龙，渡过难劫就可以化为真龙的一种，拥有强大的力量，象征着华夏强大的生命力和白衣天使，舍生忘死、逆行而上的英雄主义精神，华夏大地在疫情后慢慢地苏醒，像是沉睡的蛟龙，在苏醒那刻将会震惊世界，所有在沉寂中积蓄的能量，必将会迎来最绚烂的绽放。

Dragons are Like People
Silk, organza, imitation fur...

蛟似人
丝绸、欧根纱、仿皮毛材质……

Huang Gang

黄刚

This series of clothing design inspiration comes from the *Dapeng Wings to run "dream"* propaganda works. Shenzhen district known as "Pengcheng", Dapeng exhibition super metaphor meaning Shenzhen in the spring breeze of reform and opening up to fly, brave wave head to run "dream", this dream is the Chinese dream, the world dream, the people's dream. The yarn of the design works adopts environmental protection recycled polyester fiber certified by GRS and the invention patent BULKY WARM volcanic high strength thermal storage fiber, organically integrating the traditional Jacquard pattern of the East and the classic dry bird grid pattern of the West. Using Chinese Cixing knitting computer flat knitting machine imitation embroidery technology to weave fashion patterns, environmental regeneration and science and technology materials into personalized knitted garments, so that the eastern culture and western fashion organic combination, let the Chinese dream fly to every corner of the world.

Cultural Continuity-the Chinese Dream
Yarn using GRS certification of environmental protection recycled polyester fiber, invention patent BULKY WARM volcanic high strength thermal storage fiber

本系列服装设计灵感来源于《大鹏展翅奔"梦"去》宣传作品。深圳称"鹏城"，大鹏展超喻意深圳在改革开放的春风里展翅飞翔，勇立潮头奔"梦"去，这个梦正是中国梦、世界梦、人民梦。设计作品纱线采用获得GRS认证的环保再生聚酯纤维、发明专利BULKY WARM火山岩高强蓄热力纤维，将东方传统提花纹样与西方经典的千鸟格纹样有机融合在一起，运用中国慈星针织电脑横机仿绣花工艺，将时尚图案、环保再生与科技材料编织成个性化针织成衣，让东方文化与西方时尚有机结合，让中国梦飞向世界每个角落。

文脉承续——中国梦
纱线采用获得 GRS 认证的环保再生聚酯纤维、发明专利 BULKY WARM 火山岩高强蓄热力纤维

Huang Siyun

黄斯赟

*The Taiping* is a collection of artificial intelligent interactive fashion that combines fashion, textile art, and human-computer interaction, focusing on the art of fashion with the intervention of artificial intelligence technology and attempting to build an intelligent system/method of human-object interaction.

The work draws its narrative from one of the Chinese folk arts, the Lingnan lion dance culture (in the old days the lion dance was known as 'Taiping music' which is an expression of the people's desire to exorcise evil and to live in peace and prosperity.), and uses data based on the collection and coding of the viewer's behaviour to induce physical changes in the fabric/fashion to intentionally reproduce the beauty of the lion dance's movements and rhythms. At the same time, the embodied engagement of the body stimulates the expansion of sensory and emotional perceptions triggering the integrity of the work. Based on the bodily perceptions and sensory experiences of the self-quarantine during the pandemic, which were uncontrollably amplified by information technology, the designer siyun use form transformation as a tool for communication and expression in order to discusses the issue of fashion as an intelligent 'artificial life' to rethink and reshape the dialectic relationship between interconnectedness-symbiosis in the post-pandemic era.

The Taiping
Data, nylon, mesh

《太平》是一组集合时装纺织品艺术和人机交互的智能交互时装，聚焦于人工智能技术介入的时装艺术且试图构建一个联姻人类－万物的智能交互系统／方式。

作品从中国民间艺术之一－岭南舞狮文化（旧时狮舞称"太平乐"，寄托了人们渴望驱魔除恶，太平盛世的美好愿景）中展开叙事，利用基于采集与编码观看者的行为数据诱发织物／时装产生物理形变，意向性地再现狮舞动势和韵律之美。与此同时，身体的具身化参与，激发拓展感官与情感的认知触发了作品完整性。作者基于疫情时期的居家隔离空间中被信息技术肆意放大的身体认知和感官体验，设计师 siyun 将形式转换作为交流和表达的工具，以讨论"时装作为智能'人造生命'"的议题，展开对后疫情时代重塑万物互联／共生的思辨。

太平
数据、尼龙、网纱

Johanna Braitbart (France)　　　　　　　　　　乔安娜·布拉特巴特（法国）

The world of nature and fairy tales in poetic vision of the enchanted forest. Very soft and sweet colors in an enchanted world, a dreamy vision of life. chesplate , headpiece, belts embroidered with lace beads feathers stars, butterlys and birds.

大自然的世界和童话故事中充满了诗情画意的魔法森林。在一个迷人的世界里有着非常柔和甜蜜的色彩，那是一个梦幻的生活愿景。上面装饰有蕾丝花边、羽毛、星星、蝴蝶和鸟的饰物、饰带。

Enchanted Forest
Lace, embroideries, feathers

魔法森林
蕾丝、刺绣、羽毛

Kathrin von Rechenberg (Germany)　　　　凯瑟琳·冯·瑞星博（德国）

薯莨是一种薯蓣科植物，数千年来先辈们一直用它来鞣制皮革，鞣制渔网。清朝时期，薯莨染制香云纱（莨纱、莨绸）技术已经普遍。薯莨汁这种特殊的单宁酸染料的特性让我着迷，它的特性也可以在其他红单宁染料中找到，如涩柿子和角果木。

染色液不仅可以染出漂亮的红色和棕色，而且可以使丝绸材料变得硬挺而有光泽。染料可以反复使用，随着时间的推移颜色会变深。染色织物也一样：颜色的深度和光泽随着时间的推移而增加。

香云纱（莨纱、莨绸）的面料经过反复的染色和晒干，表面覆盖着河泥，形成了典型的黑色光泽。这是最后一步，我作品中使用了这种传统技术的整个过程。

香云纱（莨纱、莨绸）的染色过程对我来说代表了一种整体的生活方式。它完全取决于自然和时间，不能被人类的意志加速。现实的时代缺乏对自然的尊重。我们想要控制自然，反而忘记了我们需要依靠它。

我的真丝作品用悬挂漂浮的形式它展示在我们的头顶，提示着我们尊重自然。

Shuliang (Dioscorea Cirrhosa) is a tuber which has been used for thousands of years to tan leather and to treat fishing nets and sails for longer preservation. The yam was commonly used to dye fabrics in the south of China, during Qing Dynasty the technique was refined to create what we know today as XiangYunSha or liang chou. The surface of the shuliang dyed fabric is treated with river silt. I am fascinated by the features of this very special tannin dye which has properties that can also be found with other red tannin dyes like astringent persimmon and a type of mangrove Ceriops Tagal. The dyeing solution not only dyes beautiful shades of reds but also creates stiffness and shine on the silk material. The dye can be repeatedly used and becomes darker with time. Same for the dyed fabric the depth of the colour and its lustre increases with time.

For XiangYunSha or liang chou the repeatedly dyed and dried fabric is covered with mud to create the typical shiny black surface. This final step I partially employ to show the whole process of this traditional technique.

The XiangYunSha or liang chou dyeing process represents for me a holistic approach of life. It completely depends on nature and time and cannot be accelerated by the will of the human being. Actual times are lacking respect for nature. We want to rule nature and forget that we are depending on her.

I use of the silk elevated floating above us as a reminder to respect nature.

Soul Silk
Silk, gambiered silk

丝惟
真丝、香云纱（莨纱、莨绸）

Jiang Kinor / Peng Qingxin / Li Aishu (Hongkong, China)　姜绥祥 / 彭青歆 / 李爱舒（中国香港）

The surface of gambiered silk fabric coated with metal alloy by using sputtering technology to produce iridescent color effects. The characteristics of traditional material and developed finishing process are presented together on the clothing. Three garments in this collection are made by coating gambiered Canton gauze with a variety of different metals. Produced in Shunde, China, the gambiered Canton gauze is dark in colour. The silky lustre and delicate jacquard surface become interlaced with black, steel blue, fold and bronze colours from the sputter coating process. The garments skifully apply the modern line of the silhouettes. The spirit and strength of the material transcend is own space into the physical surroundings.

在香云纱表面镀覆金属，产生变幻的光色，传统与现代技术结合的特征共同呈现在服装上。本系列的三套服装是使用经物理沉积技术镀覆多种金属的香云纱制成的。顺德产的香云纱颜色慎重、有丝质的光泽和细腻的提花暗纹，经过金属处理被镀覆成间蓝、间金、间青。服装运用具有现代感的直线条廓型，使得物料本身的刚毅和力量延伸到作品之外的空间。

Floating Silk 1/2
Gambiered silk, alloy

流纱 1/2
香云纱、合金

Lan Tian

兰天

"Pine wall, Daiwa horse head wall, fat beam and thin column inner patio, high-rise building and deep courtyard with small windows." This work is inspired by the ancient architecture of Lingnan that has been learned as the "Hometown of National Treasure", and combines the ancient architectural culture of Qiaoxi with clothing design. Integrating design language and fashion elements, it deeply interprets the ancient architectural culture of Qiaoxi, allowing traditional Chinese architectural culture and clothing fashion elements to collide, bearing the traces of the years, and writing stories of time. Through this work, it is expressed that the ancient buildings of Qiaoxi are hidden in the mist and rain, and look like a thousand-year-old watch.

"松墙黛瓦马头墙，肥梁瘦柱内天井，高楼深院小窗户。"此作品以被称为"国宝之乡"的岭南古建筑为灵感，将桥西古建筑文化与服装设计相结合，融入设计语言和时尚元素，对桥西古建筑文化进行深度诠释，让中国传统建筑文化与服装时尚元素碰撞，承载岁月的痕迹，写满时光的故事。通过此作品表达出桥西古建筑隐于烟雨迷萧中，恍若千年一空的守望。

White Wall Tiles
Cotton, PU, polyester, organza

粉墙黛瓦
棉、PU、聚酯纤维、欧根纱

Li Aihong                                    李艾虹

*All the way warm* works inspired by the new thinking of traditional activation design, the collision of traditional Dong handmade cotton cloth and modern technology produces new sparks, in the design of "none" design as the ultimate goal of design, with the number one as an individual, continuous development and integration into a force, it seems that the clothing presentation of cohesion expression, only through the transformation of a piece of Dong cloth fabric. However, they can feel the possibility of wearing curvature changes and the convenience of wearing free changes, which makes it possible to design fashionable and futuristic clothes. Use design to warm people's hearts, use design to care for the nation, use design to convey Oriental philosophy of life, and convey the power of Oriental fashion creativity.

《一路温暖》作品灵感源自对传统活化设计的新思考，对传统的侗族手工棉布与现代技术碰撞产生新的火花，在设计上以"无"设计为设计的终极目标，用数字一作为个体，不断的发展集合成一种力量，似乎对凝聚力表达方式的服装呈现，仅仅通过对一块侗布面料的改造，却从中感受曲率变化的穿着可能性与穿着自由变化的便捷性，使设计出时尚未来感的服装成为可能。用设计温暖人心，用设计关爱民族，用设计传递东方人的生活哲学，传递东方时尚创意的力量。

Warm All the Way
Cotton, Silk

一路温暖
棉布、丝绸

Lee Yeonhee (Korea)　　　　　　李莲姬（韩国）

The design was inspired by an image of artic ice broken off from a nearby glacier and melting in the heating ocean. Using a hand bleach approach on the denim fabric and the modular design concept, the designer aimed to highlight both the sublime beauty and the shifting condition of the planet. The method of using 3D prints as modules that connected with denim fabric through 3D printed buttons was a new way of incorporating technology into modular designs as well as increase the durability of the modules.

该设计的灵感来自冰川断裂并在加热的海洋中融化的北极冰的图像。设计师在牛仔布面料上使用漂白方法和模块化服饰设计理念，旨在突出地球的崇高之美和全球变暖的趋势。设计运用了3D打印机打印出来的模块和按钮与牛仔布连接的方法，将传统与现代技术融入模块化的服饰设计，并增加模块服饰的耐用性。

Wish for Happiness and Wellness
Yarn knitting

唯愿幸福安康
纱线　针织

The openness or liberation we have sought for a long time is essentially to better protect women's bodies and rights. The work is inspired by the biological defensive shell generating structure, starting from the concept of "defense", using European geometry and Euclidean geometry as element variants, and recreating the mathematical aesthetics through 3D modeling, variant parameter design and other means. Explore clothing to shape the human body, create a virtual "re-incarnation", seek women's exploration needs of the virtual and real world, and achieve introspection and relief.

长久以来我们寻求的开放或者解放本质上是为了更好地保护女性的身体与权益。作品以生物防御性的壳体生成结构为灵感，从"防御"的概念出发，以欧式几何和非欧几何为元素变体，通过3D建模、参数化设计等手段再现重构数理美学。探索服装对人体的塑型，创造虚拟的"再化身"，寻求女性对虚拟与现实世界的需求，实现反思和解脱。

Li Yang / Qian Jinyu

李洋 / 钱瑾瑜

Defense
Cowhide, silk and hemp compound chiffon

防御
生牛皮、丝麻复合雪纺

Li Wei

李薇

*The Youth of Jinse* is inspired by the poem *the Sad Zither (Jinse)* by Li Shangyin, the poet of Tang dynasty: 'Why should the sad zither have fifty strings? Each string, each strain evokes but vanished springs'. It is a metaphor for the youth age, and it also implies the arrival and bright future of Song brocade. Through the continuous exploration of Song brocade's traditional and artistic value, and more diverse and younger fashion forms, this artwork highlights the commercialization, artistry and internationalization of Song brocade fashion, and integrates traditional culture into international fashion trend, moving towards future and the international arena. Let Song brocade be like a young child, glowing with new energy and vitality, and relive the great moments of youth.

灵感来源于唐代诗人李商隐《锦瑟》一诗："锦瑟无端五十弦，一弦一柱思华年"，比喻青春时代，也寓意着宋锦美好时代的到来和未来发展。通过不断挖掘宋锦传统及艺术价值和更多元、更年轻的时尚表现形式，使本件作品凸显宋锦时尚的商业化、艺术化、国际化，让传统文化融入国际时尚，走向未来、走向世界。让宋锦像一个年轻的孩子一样，焕发出新的能量与活力，重温锦瑟华年。

The Zither
Song brocade, raw silk, satin

锦瑟
宋锦、生丝、绡

Li Yiwen

李怡文

The work *Fold the continuation of embroidery* is inspired by the small triangles in the overlapping embroidery (also known as "barbola") of the Miao ethnic group. The superposition and accumulation of each small triangle represents the accumulation of life experiences, whether good or bad, sour, sweet, bitter or spicy, which are all part of the growth of life. Laser cutting, hand-made, flat cutting combination; In terms of color, the main fabric uses grey green leather, giving a quiet, thinking mood; Production of laminated fabric, choose stiffer metallic luster fabric, with a sense of modern; Gold and silver represent gold and silver in ethnic minorities, which means to be handed down from generation to generation. This group of works is presented in the form of concept clothing with the combination of handmade temperature sense and modern art sense, and the small triangle in the work can be used continuously or can be changed into various shapes. It has sustainability and development, telling the endless vitality.

Fold the Continuation of Embroidery
Leather, rope, metallic yarn

《叠续》这组作品是从苗族叠绣(也叫"堆绣")中一个个小三角形中提取的灵感，每个小三角的叠加、堆积，都代表着人生阅历的积累，无论是好的、坏的、酸的、甜的、苦的、辣的，都是人生成长的一部分。工艺上采用了激光切割、手工制作、平裁、立裁相结合；色彩上，主面料用了灰绿色的皮革，给人安静、思考的情绪；制作叠绣的面料，选用较硬挺的金属光泽感面料，具有现代感；而金银色在少数民族中又代表着黄金、白银，寓意为子孙代代发财。这组作品以手工的温度感与现代艺术感结合，以概念服装形式呈现，且作品中的小三角可以持续使用，也可以变化出各种造型，它具有可持续性和发展性，诉说着源源不断的生命力。

叠续
皮革、绳子、金属纱

Li Yingjun                                    李迎军

The work focuses on the manifestation of Dunhuang art in modern design from the perspective of internationalization, and tries to explore the international value of Dunhuang art through design practice.

作品关注国际化视角下的敦煌艺术在现代设计中的显现形式，努力通过设计实践探究敦煌艺术的国际化价值。

Apsara
Cotton, wool ,silk

御风
棉、毛、丝

Debbie Leung (Hongkong, China)                    梁楚茵（中国香港）

The Solar Eclipse
Was it night in day?
When nature has its say
Everyone ought to obey
It has to be that way.
Vikram G. Aarella

日食
是白天的夜晚吗？
当大自然有了发言权
每个人都应该服从
必须是这样。
维克拉姆 · 阿雷拉

Eclipse
Handmade felt with wool and silk

圆缺
手工毛毡、羊毛、丝绸

Liang Li / Hou Sijia

梁莉 / 侯思嘉

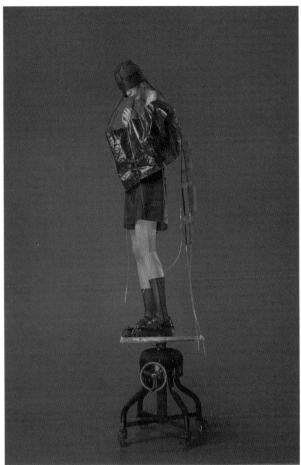

It's a dialogue among human clothing and the environment.
Gambiered Canton Gauze is soft while Dong cloth is stiff. Both of them are coated fabrics made by traditional techniques. The composition of the coating is used to protect the human body from the external environment.
Protection, half comes from external power, and half comes from inner power. The form of protective clothing brings more power to the heart, and the inner power is the final protection of mankind.

这是一场人、衣、环境之间的对话。
香云纱和侗布，一柔一刚，但都是传统工艺制作的涂层面料，涂层的成分用于保护人体免受外部环境的侵害。
保护，一半来自外部力量，一半来自内心力量。防护服的形式给内心带来更多力量，而内心的力量是人类的终极保护。

Final Protection
Gambiered canton gauze, dong cloth

终极保护
香云纱、侗布

Liang Mingyu                              梁明玉

Discarded things can also be turned into beautiful fashion. *Eco-jeans* is based on inventory, waste jeans and data lines as the main materials, with unrestrained creativity, subverting the traditional concept of ready-made clothes, as far as possible to maintain the characteristics and texture of jeans, giving a new structure and space, all over the body of the data line, showing the spiritual landscape and lifestyle of contemporary youth. By deconstructing the aesthetic stereotype of jeans, the world's most carbon-intensive product, I hope to give you a warning to tap the stock reduction increment and a declaration of environmental protection.

废弃的东西也可以变得美丽、时尚。《生态牛仔》是以库存、废弃牛仔裤、数据线为主要材料，用天马行空的创意，颠覆了传统的成衣概念，尽可能保持牛仔裤的特性和肌理，赋予全新的结构和空间，遍布全身的数据线，呈现着当代青年的精神景观与生活方式。
通过解构牛仔裤这种全球碳排放量最大产品的美学刻板印象，希望给大家予于挖掘存量、减少增量的警世劝告和环保的宣言。

Ecological Jeans
Jeans cotton, data line, PU tube, soft iron wire for textile

生态牛仔
牛仔裤棉布、数据线、PU 管、纺织品用软铁丝

In the connotation of oriental philosophy, the idea of "harmony between human and nature" is contained in the intangible temperament and charm, which can be embodied among the stretching mountains and waters. Chinese literati landscape painting perfectly blends the image of heaven with the will of human, and the self-spirit of artists reappears under the representation of natural landscape. The brush painting of China melts with water, with appropriate shades and multifarious postures. The delicate brush strokes are blooming in the ethereal brush paintings, slightly shady and gestated in the implication of "The great sound seems soundless, the great image seems formless ", as if returned to the original state of ecstasy, the so-called dream is not a dream, real is not a real, everything is nothing…

*A Dream of Floating Life* is intended to express that the ups and downs of life are like dreams, ethereal and unable to control. Life is both true and fictional, both virtual and real, like a dream in the work. "The floating life is like a dream, while the floating dust is like nothing." The twisted connection between human and the world makes us be dominated by fate and drift along the stream. The original intention has been covered up by the secular world and become more and more confused. We can only review and retrospect ourselves more vividly by breaking through the fog to discover the original self and return to the original dream.

Liang Zhiyin                    梁之茵

在东方哲学内涵中，"天人合一"的思想观念蕴含在无形的气质与神韵中，在绵延的山水之间得以体现。中国文人山水画将天的物象与人的意志完美融合，在自然山水的表象下，艺术家的自我精神再现。中国水墨遇水则融，浓淡相宜、姿态万千，细腻的绘画笔触在飘逸的水墨晕染中绽开，若隐若现，孕育于"大音希声，大象无形"的意蕴中。仿佛如回归初心的忘我之境，所谓梦非梦，实非实，皆虚无也……

《浮生一梦》意在表达人一生的沉浮宛如梦境一般，缥缈而不能自已。它亦真亦假、亦虚亦实，如作品一般恍若梦境。"浮生若梦，而浮尘若空。"我们与世界盘综错节的联结，让我们被命运主宰而随波逐流。初心被世俗所掩盖，而愈发迷惘。拨开层层迷雾，才能更加真切地审视与反思，发现最初的自己，回归最初的梦想。

Floating in Dream
Silk

浮生一梦
丝绸

Liu Jinyu

刘锦玉

Inspired by the ancient Silk Road, the work uses only cotton rope as the main material for weaving, winding, embroidery, printing and dyeing of different structures, and then makes it according to the pattern textures of the maritime Silk Road and mountain with drifting clouds Silk Road, hoping to express the material, space, texture and meaning perfectly. Through patient and delicate manual work, comparative accumulation and change, combined with stitching, the soft and loose cotton rope material has a heavy feeling being like stone in relief and a vigorous momentum being like breaking through brambles and thorns.

The author boldly organizes and presents his cognitive views on the Silk Road and communication, life and nature in the way of reconstructing texture with patterns and materials, and integrates artistic aesthetics and weaving techniques to form a gorgeous and unique soft sculpture being like picture.

作品以古老的丝绸之路为灵感，整体采用单一的棉绳线为主材质，进行不同结构的编织、缠绕、绣绘、印染，再依据海上丝路与云山丝路的图案骨骼展开制作，希望把材料、空间、质感、意义等表达得尽善尽美。通过耐心、细腻的手工劳作，对比堆积及变化，配合缝缀针法，使柔软松散的棉绳线具有了石浮雕般的沉重感与披荆斩棘般的雄浑气势。

作者大胆地将自己对丝路与交流、生命与自然的认知观点用图案与材料再造肌理的方式来组织和呈现，将艺术审美与编织技法相互融合，形成绚烂又别致的软雕塑般的画面。

Description of the Silk Road Clouds Picture
Cotton thread, knitting, yarn and silk

丝路云图
棉线、针织、纱绸

Liu Jin                                        刘瑾

Clothes, like masks, cover the body, but expose our living state and mood. At present, there is always a delicate relationship between life and nature, and between people. The seemingly transparent material is cold, the sharp tentacles are repellent but fragile. Warm fragrant cloud yarn connects the outer world with the inner world.

衣服如面具，遮掩着身体，却裸露着我们的生存状态与心境。在当下，生命与自然之间、人与人之间，总是存在微妙的关系。看似透明的材质却冰冷，尖锐的触角拒人千里却又脆弱。有温度的香云纱连接外部世界与内心。

State • the Mask
Matal wire, gambiered canton gauze, polyster

状态·面具
金属丝、香云纱、聚酯材料

Liu Jing                                        刘静

This series is called *Ning*. I want to show people that the sustainable fashion isn't just about simple organic linen or some biodegradable materials, but it can be a highfashionable and romantic item. I realised the huge challenges of fashion industry when I started my research about sustainable design. The fashion industry is the world's second-largest pollution industry after the oil industry. Therefore, inspired by the deconstruction of architecture, this series uses block surface segmentation to transform and upgrade the abandoned clothing and fabrics into new garments that can be worn in a variety of ways. Dark purple is widely used in the design to combine with hollowing, splicing and other elements to explore the space between deconstruction and the human body as well as the rock style. Soft gray-purple and bright green with high saturation are injected into it, expressing the collision of rock and romance.

本系列名为《凝》，我想通过这一系列的设计告诉人们，可持续时尚不仅只是朴素的有机亚麻或者一些可降解的材料，也可以是高街且浪漫的时尚单品。当我开始可持续时尚研究以后，在行业中的实践调查和文献数据让我越发意识到时尚产业的浪费问题，它是仅次于石油产业的第二大污染产业。因此这一创作以解构建筑为灵感，运用块面分割将废弃的服装及面料重新组合，变化成以多种方式进行穿搭的服装款式。服装色彩以深紫色调为主，通过与镂空、拼接等元素结合，探索解构主义与人体之间空间的同时，也营造更强的摇滚风格。柔和的灰紫色以及具有高饱和度的明绿色注入其中，通过色彩表达摇滚与浪漫的对碰。

Ning
Cotton thread, knitting, yarn and silk

凝
棉线、针织、纱线和丝绸

Liu Jun 刘君

The work *Decorated by Green* is inspired by the spring green outside the window in April. In this work, firstly, combined with the traditional craftsmanship of Miao nationality-applique and superposed embroidery, the ethnic minorities' traditional handcraft is transformed, interpreted and innovated, and the tender green fabrics are cut into leaf patterns in the form of applique to embroider on dark green organza. Secondly, the two-color organza is embroidered layer by layer with superposed embroidery to form a rich pattern on waist part. The design of this work not only contains the essence of ethnic minorities traditional handcrafts work, but also combines with the style of modern fashion.

《缀绿》的创作灵感来源于四月窗外的春绿，作品结合苗族传统手工工艺贴花和叠绣，将民族手工艺转换演绎与创新，运用贴花形式将嫩绿色面料剪成树叶图样，拼连贴绣于墨绿欧根纱上；运用叠绣将两色欧根纱层层堆绣而成变化丰富的腰部图案，作品设计既蕴含民族传统手工艺精髓，又与现代时尚相结合。

Green

Organza

缀绿

欧根纱

The work is inspired by the color elements of opera and lion dance culture in Xiyi characteristics. The strong regional characteristics give people a strong visual impact. The rich colors well explain the post epidemic era. Although mankind is facing various difficulties, it still lives a vibrant life and still lives in the sun.

The traditional silk gauze and denim fabric are combined, and the overall shape is based on the circle, which expresses a harmonious "symbiosis" concept in the conflict of concave convex, soft and hard.

作品灵感来源于西樵特色中戏曲和舞狮文化的色彩元素，富有浓郁的地域特色，给人很强的视觉冲击力。丰富的色彩，很好地诠释了后疫情时代，人类虽面临各种困境，却依然把生活过的生机勃勃，依然向阳而生。

将传统真丝绡和牛仔面料相结合，整体造型以圆为基本，在凹凸、软硬的冲突中，表达一种和谐"共生"理念。

Liu Na

刘娜

108

Symbiosis
Silk, denim

共生
真丝绡、牛仔布

Liu Wei

刘毽

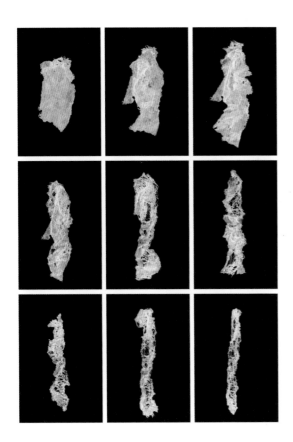

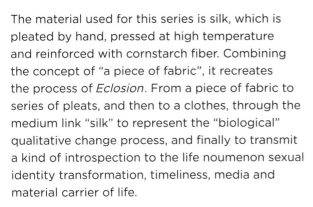

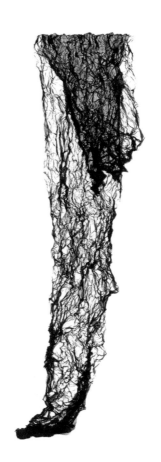

The material used for this series is silk, which is pleated by hand, pressed at high temperature and reinforced with cornstarch fiber. Combining the concept of "a piece of fabric", it recreates the process of *Eclosion*. From a piece of fabric to series of pleats, and then to a clothes, through the medium link "silk" to represent the "biological" qualitative change process, and finally to transmit a kind of introspection to the life noumenon sexual identity transformation, timeliness, media and material carrier of life.

《晓梦迷蝶》系列使用材料主要是真丝绡，通过手工制褶、高温热压与玉米淀粉纤维增强塑形的工艺，结合"一块布"的理念，以重现"羽化"的过程。从一方布，到一席褶，再到一件衣，通过"丝"这种物质媒介的链接，是对蚕蛹吐丝作缚为茧、破茧羽化振翅的"生物性"质变过程的模仿与再现，以表现对于生命本体身份性的转换，对生命时效性、媒介性和物质性载体的思考。

Eclosion

Silk

晓梦迷蝶

真丝绡

Rosew Liu                                    刘薇

The starry sky that cannot be swallowed by darkness is condensed with a force that transcends nature, a kind of profoundness that is full of unknown. It shows different postures in different times, spaces and environments, and it continuously gives us endless inspiration. However, increasingly serious environmental problems are destroying our homes for survival. Let us pay attention to environmental protection, watch the stars, and watch the unknown future.

In the era of material surplus, how to create and think in a more environmentally friendly way. In the process of making clothing, 20%~30% of the fabric is leftover material from cutting, which is inevitably "abandoned". This is especially true in the high-level customization process of pursuing the ultimate beauty. The creative point of this group of art installation works hopes to advocate for more people to pay attention to the concept of environmental protection, sustainable regeneration, and recycling.

Watch the Starry Sky
Fabric

无法被黑暗吞噬的星空凝聚着一种超越自然的力量、一种充满未知的深刻。它在不同的时空、不同的环境下展现着不一样的姿态，源源不断地给予着我们无穷的灵感。然而日益严重的环境问题正在破坏我们赖以生存的家园，让我们一起关注环保，守望星空，守望未知的未来。

在物质过剩的时代，如何用更环保的方式去创作去思考。在服装的制作过程中，20%～30%的面料是裁剪的剩余材料，不可避免地被"遗弃"。在追求极致美的高级定制过程中更是这样。这组艺术装置作品的创意点希望倡导更多的人关注环保，可持续再生，循环再利用的理念。

守望星空
服装面料

Liu Xiangqun                                        刘骧群

Iris Van Herpen is a famous Dutch Fashion Designer, whose clothing created by 3D printing technology has overturned our cognition and definition of fashion design. She combines the most traditional handicraft skills with the latest high-tech technology and materials in the real world, and creatively realises the perfect integration of the two worlds. Correspondingly, this work is made with Chinese traditional paper-cut, paper binding, stitching, paper lamp and other technology, aiming to imitate and look at the high-tech 3D printing technology in the creative form language through Chinese traditional manual technology, while homage to Iris Van Herpen, it is also exploring the way of innovation and the innovation between Chinese traditional culture and traditional technology to modern high-tech technology, and how to move towards international. It not only expresses the integration of the past, the present and the future, but also shows and extreme, contradiction, reshaping the context of the historical revival and the present.

Iris Van Herpen是荷兰著名服装设计师，她用3D打印技术创作的服装彻底颠覆了我们对服装设计的固有认知与定义。她将传统的古老手工艺与现实时代最新的高科技技术、材料相结合，创造性地实现了两个世界的完美融合。与此相对应的是，本作品采用了中国传统剪纸、纸扎、刺缝、纸灯等工艺技术，旨在通过中国传统的手工技术在创作形式语言上来模仿、遥望高科技3D打印技术，同时向Iris Van Herpen致敬，也在探索中国传统文化与传统工艺如何与现代高新技术结合、创新的道路，如何走向国际，在表达过去、现代及未来的融合同时，也展示了一种极致的矛盾，重塑了历史复兴与现实当下的语境。

Salute
Paper, cane

致敬
拷贝纸、藤条

For young people with fast pace of life and time fragmentation, they will need to consider more spiritual needs. The art trend of printing adds some post-modern design concepts and cultural elements to realize people's spiritual needs and emotional sustenance, and make some embellishment fun for impetuous and fast-paced life.

The round head is like a reflective pearl, which also represents every girl who is unique as a pearl and treasure; It is also like wishing a magic ball to reveal the inner world of every girl and pray for the magic ball to come true. At the same time, the round shape arouses the girl's desire to hug, hoping to give a warm hug to the lonely soul.

对于当下生活节奏快、时间碎片化的年轻人而言，会需要更多地考虑精神需求。印花品的艺术潮玩，加入一些后现代化的设计理念和文化元素，去实现人们的精神需求和情感寄托，给浮躁、快节奏的生活增添一些点缀的乐趣。圆圆的脑袋似反光的珍珠，也代表如珠如宝独一无二的每一个女孩；也如许愿魔法球，揭示每一个女孩的内心世界所愿并想魔法球祈祷实现。同时，圆润的造型激起少女想要抱抱的欲望，希望给孤独的灵魂一个温暖的拥抱。

Lau Sanchia (Macao, China)　　　　　　刘欣珏（中国澳门）

Chiwawa
Canvas

痴娃娃
油画布

Liu Xun

刘寻

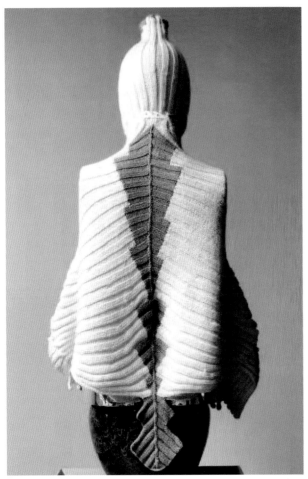

2020 is a special year, no matter who experiences anything, all want to fly to the brave bird.

2020年是非凡之年，无论任何人经历任何事，都要像勇敢飞行的鸟……

Flying Bird
Wool

飞行鸟
羊毛

Liu Yang                                                   刘阳

The texture reflected from the inheritance of Dong fabric technique from Dong nationality alternates with the fashionable and shining sequin fabric, as the aspirations and pursuit of Dong people to live a better life are just brewing up in its natural shine, colors in various lights, simple and heavy natural texture, traditional technological process. The characteristics of both materials are fully considered in their combination and utilization. Designs are a mix of the characteristics of clothing from the East and the West, as the natural overhanging or tying way of wearing with upper outer garment and lower hem is suitable for the seriousness or casualty and ease in different occasions.

侗族的侗布工艺传承体现出的质感与亮片布的时尚闪耀交相辉映，其自然光泽感、不同光线下的色泽与质朴厚重的天然质感，传统的工艺流程，酝酿着侗族人民对生活的美好向往与追求。在两种材质的结合与运用上，充分考虑了其特点。款式上融合了东西方服饰特点，上衣下摆自然悬垂或系结的穿法，适合严肃或随意自然的不同场合。

Rhythm of Dong Minority
Fabric of dong minority, light piece cloth

侗韵
侗布与亮片布

Liu Yihang                              刘一行

What I want to express is that a person's space has no size, and he carries more things. Today's fast-paced life wants to find his own comfortable space.

我想表达的是：人的空间没有大小，他承载的东西比较多。现今快节奏的生活想要找到让自己比较舒适的空间。

Left Hand
Polyester fiber, wool

左手
聚酯纤维、羊毛

Inspired by the changes of Chinese costume culture in the past hundred years, the classic style of collar is extracted from the tradition to the contemporary evolution, integrated with natural elements, and Joined all kinds of embroidery techniques for creation and processing to inherit the classic modern decoration. The implication of Chinese and Western cultural convection, showing the multi-cultural spirit far beyond other concrete functions and significance.

以百年间的中国服饰文化变迁为灵感，在传统到当代的沿革中提取衣领的经典款式，融入自然的造型元素，加以各式绣法进行创作处理，以现代的装饰方式传承经典。寓意中西文化对流，显示出的多重文化精神远远超出其他具象的功能和意义。

Lu Yan

陆琰

The Field
Tong cloth, embroidered fabric, beads, rhinestones, etc

领 · 域
亮布、刺绣花片、珠片、水钻等

Luo Jie                                    罗杰

The design elements are based on the wide-color porcelain which integrates the Chinese color porcelain and the western color porcelain technology. The clothing outline absorbs the basic modeling characteristics of porcelain on the basis of the classic dress.

Guangcai is the crystallization of the exchange and integration of Chinese and Western cultures. It is fired by the combination of Chinese traditional techniques and Western techniques to form a unique style with gorgeous colors, complex processes and cumulus. Therefore, from the material of clothing, the classic mesh, milk silk and other materials are digitally printed, combined with the electronic variable optical fiber textile fabric with great scientific and technological texture, showing a changeable color visual relationship. In addition, the classic paper-cut elements of dyed rice paper are added, which has the unique complex beauty of the combination of Chinese and western.

设计元素以融合了中国彩瓷与西方彩瓷技艺的广彩瓷器为原型，服装廓型在经典礼服的基础上吸收了瓷器的基本造型特征。广彩是中西文化交流融合的结晶，由中国传统技法和西洋技法结合烧制形成了色彩绚丽、工艺繁复、堆金积玉的独特风格。因此，服装从材料上将经典的网纱、牛奶丝等材质进行数码印花处理，结合了极具科技质感的电子可变光纤纺织面料，呈现出多变的色彩视觉关系。此外还加入了经典的染色宣纸的剪纸元素，有独特的中西合璧的繁复之美。

Urban Evolution
1.75mm printed TPU, hard organza, printed elastic mesh, cotton, electronic variable optical fiber textile fabric, digital printed milk silk, dyed rice paper

城市进化
1.75mm 印花 TPU、硬质欧根纱、印花弹力网纱、棉花、电子可变光纤纺织面料、数码印花牛奶丝、染色宣纸

Luo Juan                                    罗娟

Continuing the idea of the previous red culture-themed works *The Bitter Joys of The Red Years*, still adopts the techniques of deconstruction and splicing to transform the materials into the more suitable contemporary Eugen satin, which implies that the younger generation should cherish the present happy life.

延续作者红色文化主题作品《红色岁月之苦尽甘来》的立意，仍然采用解构、拼接的手法，将材料转换为更加适合当代的欧根缎，寓意是年轻一代要珍惜现在的幸福生活。

The Red Future
Organza

"红"之未来
欧根纱、欧根缎

Over the past century, human civilization and development have left a deep impression on the earth and the environment, such as water pollution, white garbage, the greenhouse effect and the melting of icebergs. We should reflect on our own actions while pursuing contemporary life.

This series uses iron embroidery dyeing and plant dyeing to explore the reflection of contemporary culture expressed in the language of "disorder and abstraction" and the language of garment structure with zero waste.

The real and the virtual are inherently interdependent, just as we are interdependent with the earth and the environment.

Luo Ying / Wu Hongcheng                                   罗莹 / 吴红程

百年来，人类的文明和发展留给地球与环境深深的印记，如水污染、白色垃圾、温室效应、冰山融化。我们在追求当代生活的同时应当反思自身的作为。

本系列用铁锈染和植物染来探讨用"无序和抽象"的语言、用零浪费的服装结构语言表达对当代文化的反思。

实与虚之间本来就相互依存，如同我们与地球、与环境相互依存的关系一样。

Printing Marks
Silk, linen

印迹
丝、麻

Cindy Luo

罗峥

The Saying of Jade is inspired by China's long-standing jade culture. The works uses different kinds of traditional silk Shunyu Qiaoqi. Silk hemp bark wrinkle, silk wool jacquard satin, Hualuo, Silk crepe de Chine etc, combining with varieties of ancient techniques such as Hot melt silk lining, hand-made nail bead embroidery, hand-made needle pile, hand-made thread pulling, hand-made gradual halo dyeing, fried tea dyeing, hand-painted and so on, shows the raw jade coming from the land naturally, and will return to the earth finally. But after a long time, they will reborn from the soil, and shine in an endless way. This free and easy, elegant works shows the deep understanding of Chinese ancient jade culture by Cindy Luo.

《璞玉说》的灵感来源于中国悠久的玉文化，作品采用真丝顺雨乔其 、丝麻树皮绉、丝毛提花缎、花罗、真丝双绉等材质，运用热熔烫丝衬、手工钉珠绣、手工针绒、手工拨丝、手工渐变晕染、炒色茶染、手绘等十多种工艺，选取了米姜黄色、东皇玉色以及岩粽绒黄色等，表现出璞玉来源于土地，未经雕琢、浑然天成的自然质朴，以及世间万物皆归于泥土，却在经历了漫长的时间、空间的变换沉淀之后，从泥土中幻化重生，并焕发出生生不息的生命力量。结合洒型自由、灵动飘逸的廓型设计，表现出当代设计师对于中国古代玉文化的深刻理解。

The Saying of Jade
Silk Shunyu qiaoqi, silk hemp bark wrinkle, silk wool jacquard satin, Hualuo, silk crepe de chine and other materials

璞玉说
真丝顺雨乔其 、丝麻树皮绉、丝毛提花缎、花罗、真丝双绉等

Lyu Yue (Aluna) / Jin Xiaoyao　　　　　　吕越 / 金小尧

A group of two works of the work *Cloud light-Grain Rain* and *Cloud light-Grain Buds* are respectively from the hand-painted works of artist Jin Xiaoyao and NFT works. Combined with the clothing structure design that fashion artist Lyu Yue is good at, the wearable fashion art works are completed by cutting and sewing through the use of digital printing technology to realize the fabrics with colors and patterns. Different types of artists learn from each other and try to explore the extensibility of painting works combined with different technologies.

《云之光—谷雨》《云之光—小满》是一组两件作品，分别来自艺术家金小尧的手绘作品和NFT作品，结合时装艺术家吕越擅长的服装结构设计，利用数码印花技术实现有色彩和图案的面料，经过裁剪缝制工艺完成可穿的时装艺术作品。不同类型的艺术家取长补短，尝试探讨绘画作品结合不同技术扩宽艺术的延展性。

Cloud Light-Grain Rain / Cloud Light-Grain Buds
Chemical fiber fabric

云之光—谷雨 / 云之光—小满
化纤面料

Lyu Yue (Aluna) / Fan Yuming                    吕越 / 范玉明

The work *New Needlework (Nv Gong)* brings out the oriental scene with bright red visual tone. It uses silk, Zhangzhou velvet satin, Shu brocade, Song brocade, Tujia brocade and Su embroidery to shape the corsets, highlighting the feminine beauty and strength. The shape of the human body and the unique cutting line present a diverse female image.

The work uses the other pronunciation of the Chinese character 红 in *Nv Gong (needlework)*, which sounds "Hong" with the meaning "red" as its visual tone.

During the exhibition, supplementing by hundreds of roses, the clothing is accompanied by the process of blooming and withering of the flowers in a collision that complements each other, witnessing the fusion of memory and reality in this handcraft and human body, and meanwhile highlighting the extensive and profound Chinese traditional textile culture.

The skills inheritor of Kesi (Silk Tapestry), Yuming Fan;

作品《新女红》以鲜红的视觉基调让东方情境跃然而出，以丝绸、漳缎、蜀锦、宋锦、土家锦以及苏绣等精巧工艺塑造紧身胸衣，凸显女性的柔美与坚强。人体形制的造型及独特的裁剪线型，呈现出多元的女性形象。

借用女红的"红"（gong）字的另一个读音红（hong）作为作品的视觉基调。

在展览中，辅以百朵玫瑰，在相得益彰的碰撞中，服装伴随鲜花盛开凋零的流逝过程，也见证了记忆与现实在这手艺与身体中形成的中和，同时也强调中国传统纺织文化的博大精深。

缂丝提供者范玉明为缂丝非物质文化遗产传承人。

New Needle Work (Nv Gong)
Silk, FRP

新女红
丝绸、玻璃钢

Lyu Yue (Aluna) / Zou Yingzi

吕越 / 邹英姿

138

The work evolved from one of Lyu Yue's representative works *Transformation* and added some contents of Suzhou embroidery by collaborating with Zou Yingzi, the Chinese arts and crafts master and inheritor of Suzhou embroidery. The whole work is composed of graphic works and three pieces of "clothes". "Clothes" is the cloth piece that is not cut or sewn, through winding and embroider frame clip completes the appearance of the clothes, which reflects the concept of "cherishing thing".

The work shows that the change of the world, the change of things, the change of nature and the change of the universe not only never stop, but also has the characteristics of mutualization.

作品由吕越的代表作《化》演化而来，增添了中国工艺美术大师、苏绣传承人邹英姿的苏绣内容。

整个作品由平面作品和三件"衣服"组成。"衣服"是没有裁剪缝制的布片，经过缠绕和绣花绷子的卡夹完成衣服的样貌，借此体现"惜物"理念。

作品表现的是世界的变化、事物的变化、自然的变化、宇宙的变化不仅永无停止之时，并且有互相转化的特质。

Mutualization
Cocoon, silk, bamboo, shiny thread, cotton thread, silk satin, silk, summer cloth, synthetic silk

互化
蚕茧、蚕丝、竹子、闪光线、棉线、真丝缎、真丝绢、夏布、合成纤维绸

Muhammad Fawad Noori (Pakistan)

穆罕默德・法瓦德・努里（巴基斯坦）

Denim fabric techniques under fashion sustainability to control waste n types with elements of typography text makes combination called *Denimography*. In this, the modern layered and dual sided outfits using denim fabric mostly along with pure printed cotton fabric by using various printing techniques based on typography theme.

《丹宁印刷术》以可持续的时尚为原则，控制牛仔面料制作过程中产生的废料，并结合文字排版的设计元素。现代的分层和双面设计，采用牛仔布料和纯棉印花面料，印花技术也多用于主题表达。

Denimography
Sustainable material , denim and cotton

丹宁印刷术
环保材料、牛仔、棉

Mu Lin                                    牟琳

The work *Jerseyplus* is a combination word, an artistic re-creation of the usual knitted sweaters and shirts. It aims to express an attitude, which is individual's art nature can be expressed by the familiar elements. Thusly, everyone is the artist, as long as he or she can awake the beauty of the internalized.

The work *Jerseyplus* shapes the balanced tension what the flowers show as their perfect blooming. This beauty of power is expressed by the common structural elements of knitted fabrics and shirt collars, which are layered by layers from large to small in an orderly or disorderly arrangement, as well as the interlaced light and shadow of different materials. The author regards arrangement with light and shadow together as brushes to depict her own artistic heart's content.

作品*Jerseyplus*是一个组合词，是对我们常见针织衫以及衬衫结构的艺术再创作。旨在表达一种艺术态度。在我们日常生活中所熟悉的元素是可以通过每个人的个性表现出其艺术的特质的。从这个角度理解，我们每个人都是艺术的创作者，只要能够将内化的心灵之美渐渐地唤醒。

作品*Jerseyplus*塑造的是花朵因绽放所展现的是一种平衡的张力之美。这种力量之美借由针织面料以及衬衫领子这一让人熟识的结构元素，通过层层叠叠、由大到小有序或无序的排列组合，以及不同的材质交错的光影来表达的。作者将排列和光影来当作画笔，尽情地描绘属于自己的艺术内涵。

Jerseyplus
Fashion fabric

Jerseyplus
时装面料

The natural silk fabric is used for plant dyeing and old-fashioned fabric re-construction, combined with 3D printing materials for garment design and creation. After composition design, various materials are arranged into cel-lular virus links and ups and downs, and show a hazy ink stacking effect. The overall color is thick and strong, with a sense of awe. It symbolizes that in the post epidemic era, after the darkness of the "virus war", people gradually see hope in the struggle and look for the truth of survival in the emptiness and reality of life.

People began to think about the relationship between human beings and nature, and the relationship between nature and science and technology. From natural materials to man-made consumables, from primitive plant manual dyeing to 3D printing technology that breaks the barriers of tradi-tional handicrafts, history is constantly changing and science and technolo-gy is developing rapidly. People need to think about whether the things we pursue have changed after climbing one technological peak after another. We should fear science and technology, but also nature. In this complex era of information, harmonious coexistence is the real way of survival.

Nong Qiongdan                                    农琼丹

运用天然真丝面料进行植物染色和做旧手法进行面料再造，结合3D打印的材料，进行服装设计创作。各种材料经过构成设计，排列成细胞病毒链接、层峦跌宕的造型，并呈现一种虚实朦胧的水墨层叠效果。整体色彩厚重浓郁，带有一种敬畏之意。象征了在后疫情时代，人们经过疫情这场"病毒战争"的黑暗之后，逐渐在斗争中看到希望的同时，在生命的虚实之中寻找着生存的真相。

人们开始思考人类与自然的关系，思考自然与科技的关系。从天然的材料到人造的耗材，从原始的植物手工染色到打破传统手工艺壁垒的3D打印技术出现，历史在不断更迭，科技在飞速发展。人们需要去思考，在翻越一座又一座技术的高峰之后，我们追求的东西是否产生过变化。我们应该敬畏科技，更应该敬畏自然，在这信息层叠的复杂的时代，和谐共处才是真正的生存之道。

Stack
Silk, gauze, fragrant cloud yarn, blended fabrics, 3D printing materials, etc.

叠
真丝绡、香云纱、混纺面料、3D 打印材料等

Pan Fan                                                潘璠

*Lotus Scene* uses denim fabric. With the unique processing techniques for denim, she used wiredrawing and sandblasting to show the effect of lotus. The skirt's shape is the deformation of lotus leaves, which sets off the charm of lotus under layers of lotus leaves. Inspired by ivory carving landscapes.

*Mountain Charm* uses silk gauze to shape various mountains and the scenery of mountains and rivers on the three-layer yarn skirt, which shows the love for the beautiful mountains and rivers of our motherland.

《荷戏》以牛仔布为面料，通过牛仔面料独特的处理手法，拉丝、喷砂等表现出荷花的效果，裙型是荷叶的变形，在层层的荷叶下衬托出荷花的妩媚动人。

《山韵》以牙雕山水为灵感，以真丝绡为材料，堆积出各种山脉的形状，在三层纱裙上组合出山川的景象，表达对祖国大好山川的向往。

Lotus Scene / Mountain Charm
Denim, silk gauze

荷戏 / 山韵
牛仔布、真丝绡

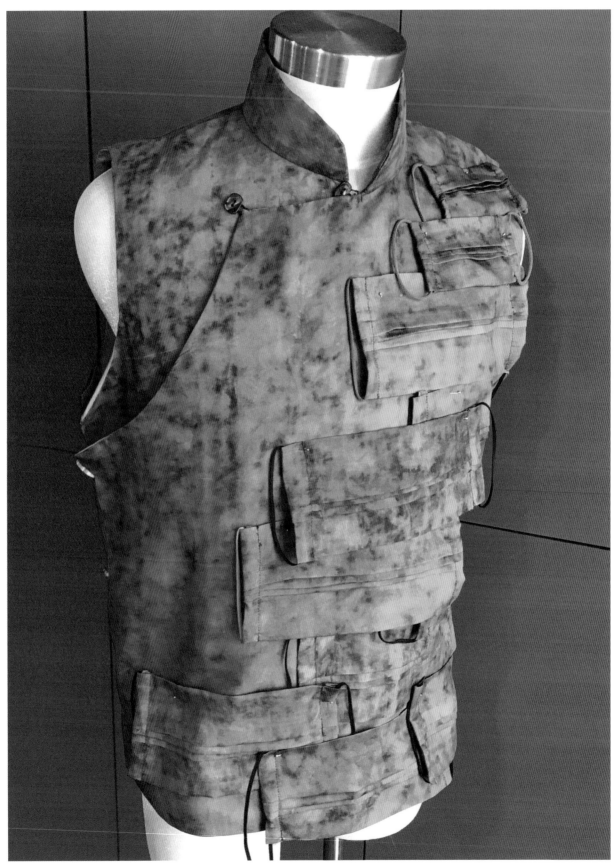

Helen Pun (Hongkong, China)　　　　　　　　　　　　潘静怡（中国香港）

*The Dress*, is a sculptural installation and the making of it is captured in a video (duration: 02:21 minutes). This artwork is a metaphoric archetype of the human body, in an involuntary state that is dependent on western medication and nutritional feed for survival. This was reflected from personal family experience with food toxin leading to aphasia and immobility. The 'dress', akin to the second skin of the body, symbolizes the helplessness state of the body–a body reinvented as a scientific construct. It seeks to evoke the fear, or perhaps a different perspective of a science-dependent physiological state of living.

This artwork attempts to interpret the relationship between Art and Science, using the body and the dress to manifest the fragility and emotions of my bed-ridden mother. Feeding bags and syringes used in this artwork were all used by my mother, as a remembrance of this journey and experience of us both.

《裙子》是一个雕塑装置，它的制作过程被拍摄在一个视频中（时长02：21分）。这件艺术品是一个隐喻的人体原型，在一个非自愿的状态，依赖于西方药物和营养饲料的生存。这反映了个人的家庭经验与食物毒素导致失语症和行动不便。"衣服"类似于身体的第二层皮肤，象征着身体的无助状态——一个被改造成科学结构的身体。它试图唤起恐惧，或者可能是一种依赖科学的生理生活状态的不同视角。

这件作品试图诠释艺术与科学之间的关系，用身体和衣服来表现我卧床不起的母亲的脆弱和情感。在这件作品中使用的喂食袋和注射器都是我母亲用过的，作为我们这次旅行和经历的纪念。

The Dress
Used feeding bags (enteral delivery systems), used syringes, colored liquids, threads, mannequin, cable ties, iron wires & mesh, tulle

裙子
用过的喂料袋（肠道输送系统），用过的注射器，有色液体，螺纹，人台，电缆扎带，铁丝网和网眼，薄纱

Qin Ruixue                              秦瑞雪

As an individual in society, people are affected by countless external influences. These influences may come from aspects such as survival, work, social interaction, appearance, etc. Each kind of anxiety gives the individual a feeling of restraint. These restraints are like different forms of nets. Wrapped around the body all the time. Weaving with industrial ties, presenting it in fashion art, show the shackles of various forms of nets on the body, and express the desire to break free of shackles through models.

人作为存在于社会的个体，受到无数外界的影响，这些影响可能来自生存、工作、社交、容貌等方面。每一种焦虑都给予个体束缚的感受，这些束缚就好像形态各异的网，无时无刻缠绕着身体。运用工业扎带进行编织，以时装艺术的方式呈现，展现出各种形态的网对身体的束缚，并通过模特表达想要挣脱束缚的期望。

Network
Industrial ties

网
工业扎带

Shao Fang                          邵芳

The theme of the series is *In the Name of Plato*. Work *Aether* is the one of this series, it uses the element of aether. It just wanted to create mysterious visual effect of future. It uses not only the structure of regular dodecahedron. The black and white rotation on the front of the skirt contains a very precise sequence relationship. It made a new design to give this dress a new visual effect. To some extent, this series works more like a teaching experiment that uses clothes to express the human-and-nature relationship and the exploration of three-dimensional space. Although these works are full of rational sequence of numbers, equalization and symmetry, but disegner still hope that design conveys the feeling of future freedom.

作品是《以柏拉图之名》系列作品之一，设计结合了中西方的哲学思想而创作。作品《以太》是这个系列之一，它使用了"以太"元素。不仅使用了规则的十二面体结构，也从音乐、数学等方面汲取了灵感。裙子的前面黑白色旋转包含一个非常精确的数列关系。使这件衣服有新的视觉效果。这样的几何系列，在某种程度上，更像是一个教学实验，用服装来表达人与自然的关系和对三维空间的探索。虽然作品充满了数字序列、均衡和对称，但设计师仍然希望设计传达出未来自由的感觉。

Aether
Knitting

以太
针织

Shi Lili 石历丽

Works in 2021 China Sanxingdui site archaeological discoveries, Sanxingdui ancient bronze mask, full pattern and three "star" form of association graphics, with PLA wire, as the 20th century greatest archaeological discoveries, shows the Yangtze river and the Yellow River basin is the mother of Chinese civilization deep culture, itself is history, nature, science, culture, art, science and technology, in the form of fashion art expression for the ancient life and the imagination of the future.

作品以2021年中国三星堆遗址考古重大发现为灵感，将三星堆古遗址的青铜面具、满饰图案及三"星"之形态的联想图形相结合，以PLA线材表现，就像三星堆遗址被称为20世纪人类最伟大的考古发现之一、昭示长江流域与黄河流域同属中华文明的母体的深厚文化含义一样，本身就是历史、自然、科学、文化、艺术、科技的溯源集结，以时装艺术的形式表达对于古老生命的回望和对未来的畅想。

Three "Star" Piles
PLA

三"星"堆
聚乳酸

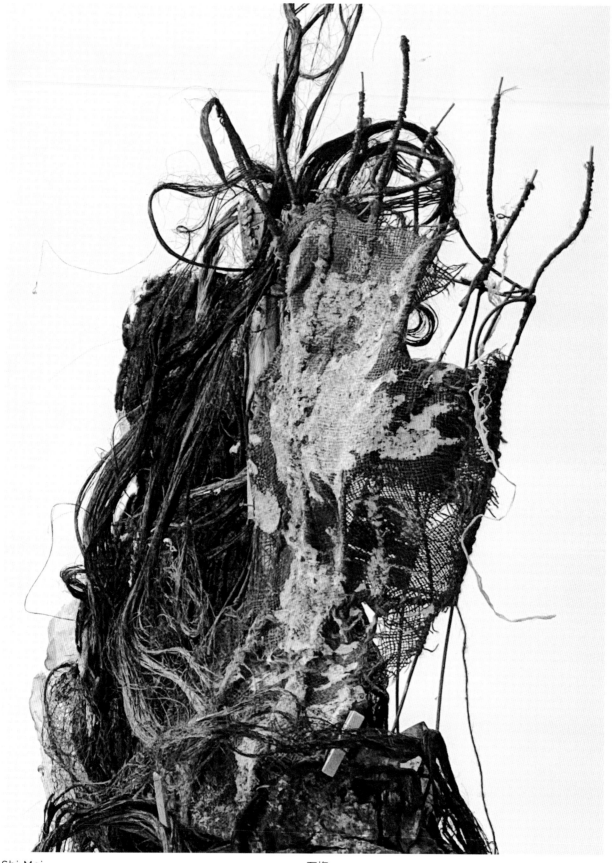

Shi Mei

石梅

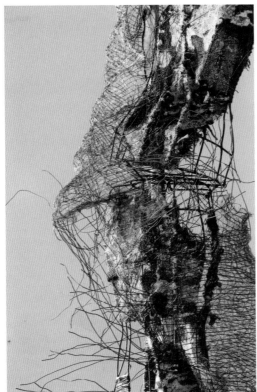

This is Debussy's orchestral work *Afternoon of a Faun*, a dreamy sound poem, which is a classic of his Impressionist music.The symphonic poem *Afternoon of a Faun* is taken from the poem of the same name by the French poet Mallarm. The shepherd in the poem has sheep horns and legs. He is the God of half man and half beast in charge of animal husbandry in Roman mythology. He grows up in the fields and is free and dissolute. The pursued fairy turns into a reed to avoid his pursuit. Since then, the shepherd takes the reed as the flute and plays it every day to express his feelings and dreams. Debussy's music is the source of inspiration for this creation. Relying on the *Afternoon of a Faun*, I express the virtual and real relationship between dream and reality. Structurally, I use two inverted and misplaced model iron frames for welding, which is also my attitude towards creation and life: all ways can be established, and there is a reasonable element in existence; In the selection of materials, iron plate, iron bar, iron mesh, worn sack rope, scorched copper wire, rusty iron wire and hemp wire are used to take the hardness, brightness and Modernity of galvanized iron plate and the ancient desolation and dilapidation of other materials, so as to make modern and ancient coexist, soft and hard virtual and real coexist and influence each other. At the same time, it also focuses on the poetic flow and ethereal ethereal of music.

这是德彪西的管弦乐作品《牧神午后》，这首梦幻般的音诗，是他印象主义音乐的经典之作。交响音诗《牧神午后》取自法国诗人马拉美的同名诗作，诗中的牧神长着羊角羊腿，是罗马神话中掌管牧业的半人半兽的神，他生长在田野间，自由自在风流放荡，被追求的仙女为躲避他的追逐化身为芦苇，从此牧神以芦苇为笛，日日吹奏，表达情思与梦幻。德彪西的乐曲是此次创作的灵感来源，我以《牧神午后》为依托，表达梦与现实的虚实关系，结构上用了两个倒置错位的模特铁架焊接，这也是我创造和生活的态度：一切方式皆有可能成立，存在即有合理成分；材料的选择上，运用铁板铁条铁网和破旧的麻袋绳子、被烧焦的铜丝、锈迹斑斑的铁丝麻线，取镀锌铁板的硬度光亮与现代感和其它材料的古旧苍凉与破败，使现代与古老共生共存，软硬虚实共存并互相融合影响，同时也更着重音乐的诗意流动与空灵飘渺。

After of a Faun
Silk fabrics, vegetable dyes, metals, wood

牧神午后
丝织品、植物染料、金属、木头

Qin Lifang                    覃莉方

The work uses flocking hot-melt adhesive film fabric for laser engraving process, and then composites with traditional mulberry silk gauze materials.The color use black white and gray to  expression ink color system.Through the combination of light laser hollowing and manual weaving, the pattern of butterfly is presented. Taking butterflies as the pattern carrier symbolizes the longing for nature.

作品使用植绒热熔胶膜进行激光雕刻工艺，再与传统桑蚕丝材料进行复合。色彩采用黑白灰的水墨色系，最终通过激光镂空与手工编织结合呈现蝴蝶的图案。运用工艺编织的手法在服装上表达中国传统水墨的意象。以蝴蝶为图案载体，象征着对自然的向往。

Ink Butterfly
Silk milk fiber flocking hot melt adhesive film fabric

墨蝶
桑蚕丝绡、牛奶丝、植绒热熔胶膜

Wang Ziding                    汪子丁

In the concept of "pathological aesthetics" proposed by the scholar Ma Weidu, the representative objects at the top of the aesthetic chain become strange or pathological due to the subjective transformation of the aesthetic subject. This is because under the monarchy society, the aesthetic development of both sexes is extremely unbalanced, resulting in the aesthetic demand for the morbid human body basically comes from men, and the aesthetic object is mainly the morbid female image. The west also has a similar aesthetic process, with women's "rich buttocks" and concave and convex curves as the beauty, are the embodiment of male aesthetic dominance. In fact, these pathological social phenomena did not exist only in ancient times. There is a similar "pathological perpetuation" in today's society-are high heels and overly skinny beauty the new age of foot binding and thin waists? When these phenomena gradually form the main aesthetic preference in society, it is also a new morbid aesthetic demand.

Based on this, I hope to use the design to call people's attention to the breeding ground of pathological beauty. As long as the society is based on this unbalanced gender construction, the pathological aesthetic will always exist. Therefore, I use in the design image isomorphism, exaggerated deformation, etc as the wrong art gimmick, selected the five of the culture in qing dynasty on behalf of the image-the goldfish, miniascape, rockery, birdcage, gall, wood and clothing isomorphism, deliberate distortion will be skewed in profile design features amplification to explore its unique artistic value, guide people by visual stimulus at the same time reflecting on the culture behind the origin, It calls for the balanced development of contemporary aesthetic standards for both sexes.

学者马未都先生提出的"病态化审美"概念中，位于审美链顶端的代表物件由于受到审美主体的主观改造，变得或形态怪异，或出现病理状态。这是由于君权社会下，两性审美发展极不平衡，导致对人体病态化的审美需求基本都来自男性，而审美的对象则主要是病态化的女性形象。西方也有类似的审美历程，以女子"丰臀"和凹凸有致的曲线为美，都是男性审美占据主导地位的体现。实际上，这些病态化的社会现象不仅仅存在于古代。在当今社会同样存在类似的"病态延续"——高跟鞋和过分追求的骨感美是不是新时代的缠足和细腰呢？当这些现象逐渐形成社会上的主要审美偏好时，又未尝不是一种新的病态化的审美需求。

基于此，我希望用设计呼吁人们关注病态美的滋生土壤。只要这个社会是基于这种不平衡的性别建构之上，病态化的审美观就会一直存在。因此，在设计中我运用形象同构、夸张变形等视错艺术手法，选取了清代文化中的五个代表意象——金鱼、盆景、假山、鸟笼、瘿木与服装同构，刻意在廓型设计上将扭曲歪斜的特征放大来探究其独特的艺术价值，引导人们受到视觉刺激的同时反思背后的文化渊源，呼吁当代两性审美标准平衡发展。

Cliche
Textile Fabric

老生常谈
纺织面料

Wang Dandan                                    王丹丹

Interpretation of the work The clothing work extracts the clothing elements of Hufu and Underwear, and expresses the abstract elements of lines and ink in Chinese ink painting through the stacking and splicing relationship of silk and satin fabrics. The ink is divided into five colors, and through the alternation of the colors of the "ink", the artistic variations of the virtual, solid, thick and light in the work are enriched. And the abstract elements of Chinese ink and wash and the harmonious artistic conception of man and nature in Taoist culture are advanced again to further convey the cultural essence of Chinese clothing and the spirit of Chinese philosophy.

该服装作品提取胡服与襦裙的服饰元素，通过真丝绡、缎面料的层叠及拼接关系，表达中国水墨中的线与墨的抽象性因素。墨分五色，通过这"墨"的色与色的交替，丰富作品虚、实、浓、淡的艺术变化。并将中国水墨的抽象性因素及道家文化中人与自然的和谐意境再次递进，进一步传达中国服饰的文化精髓与中国哲学思想的内涵精神。

Ink on the Flower
Silk satin

墨上花
真丝绡

Wang Lei                                    王雷

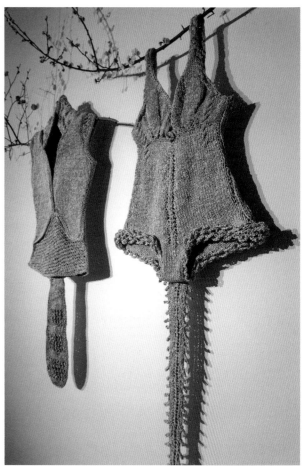

The work is based on the text of *Cihai*, drawing lessons from a part of the natural creature, such as tail, scale and other forms of language. After a tedious weaving process, the book shell part is left, which has the meaning of "ciccican peel off its shell". As a popular saying goes, "All things grow when spring thunder rings." The work uses anthropomorphic techniques to transmute a new life from the perspective of culture.

作品以《辞海》书中的内容部分，经过烦琐的编织过程，借鉴自然界生物某一部位，如尾、鳞片等形式语言，并留下书壳部分，有"金蝉脱壳"之意。民谚云："春雷响，万物长"。作品使用了拟人化手法在文化的角度上进行了一次新生命的嬗变。

Insects Awaken
*Ci hai* (chinese dictionary) paper in slubbed strings and weaving techniques

惊蛰
《辞海》纸搓线及编织

With blue dye as the carrier, there are abstract
patterns and lines of blue dye on the pattern
design. In terms of clothing style, the overall
appearance of the windbreaker is the main, and
the design of the local collar adopts the lapel and
flapper collar, combining traditional elements with
modern elements. Fabrics are made of cotton,
hemp and denim, with the concept of fashion,
comfort and environmental protection. In the
structure of the main use is the processing of the
dividing line, large surface to form a natural big
pendulum, clothing overall modeling is simple,
smooth lines.

Wang Ni

王妮

以蓝染为载体，图案设计上有蓝染的抽象图案和线条。服装款式上整体外形以风衣为主，局部领口的设计采用了对襟和翻驳领，将传统元素与现代元素结合。面料上采用了棉、麻、牛仔为主，体现时尚、舒适、环保理念。在结构上主要采用的是分割线的处理，大块面形成自然的大摆，服装整体造型简洁、线条流畅。

Blue Charm
Blue dyed cotton, hemp, jean

蓝韵
蓝染棉、麻、牛仔

Wang Peina

王培娜

The work is based on the consideration of the current natural environment, the opposition and harmony between the human world and the natural world. *The Deep End* expresses the pursuit of a free and harmonious life attitude. With the concept of environmental protection and the use of waste jeans as materials, the secondary design expresses the harmonious state between nature and human beings.

作品是基于现在自然环境的考量，人类世界与自然世界的对立与调和。《鱼翔浅底》表达了一种对自由、和谐生活态度的追求。以环保为理念，采用废旧牛仔为材质，进行二次设计来表达自然与人类之间的和谐状态。

The Deep End
Denim fabric, gambiered guangdong gauze

鱼翔浅底
牛仔、香云纱

Wang Wen 王文

This work combines new material, intelligent technology and art aesthetics to explore the new direction of the future of fashion technology. Through design and application of intelligent fiber materials, the garment can present different luminous artistic effects in different environments. At the same time, as the second skin of human body, it can help people receive the information of external environment more clearly and make corresponding responses. The work effectively promotes emotional communication and interaction, as well as health management functions.

本作品将新材料、智能科技与艺术美学相结合，探索未来服装科技新方向。通过对智能纤维材料的设计与运用，使一件服装在不同的环境下呈现灵活多变的艺术效果。同时，作为人体"第二皮肤"，智能纺织材料也可帮助人们更清晰地接收外界环境信息，与之做出相应反应。本作品有效促进情感交流与互动以及健康管理等功能。

Flow
Intelligent fiber materials

心流
智能纤维

Wang Yutao

王钰涛

Two garments were inspired from Liangzhu culture. Neither representative patterns nor dressing methods in Liangzhu culture applied in the fashion design. The designer extracted the colours and lustre of jade that was deeply buried and oxidized in the earth and could not be seen by modern people. He used round vortex patterns and Panchi stripes in Liangzhu culture and re-depict the flower, grass and other animal and plant patterns consistent with modern people's aesthetics. Moreover, he mixed and combined the colours in a popular street style, which revealed the unruly style of Liangzhu culture.

本次服装作品以良渚文化为背景，却没有直接地将良渚文化中代表的纹样图案用在服饰中，也没有将那个时代穿衣搭配方式作为依据，但提取了深埋在大地里，被氧化了现代人看不出的玉的色泽，同时应用良渚文化里圆涡纹、蟠螭纹的线条，重新描绘符合现代人的审美标的花、草等动植物图案，并通过色彩的组合与时下流行的街头混搭风带出良渚文化的不羁新风貌。

Secret Fragrace
Chiffon, silk, viscose, wool

暗香
雪纺、真丝、黏胶、羊毛

Wang Yue

王悦

The Design works of *XIE YI* used the Clamp Resist Dyeing craft in the Chinese folk traditional indigo dyeing skills, attempt to explore and innovate the traditional technologies on the transparent fabrics and carry the beauty of traditional indigo in a colorful modern way, exploring the multiple states of inner balance between tradition and modernity.

作品《颉逸》是作者与国家级非物质文化遗产南通蓝印花布传承人吴灵姝女士共同创作完成的。遵循着尊重工艺的精神，设计中通过使用透明材料对传统蓝染工艺进行色彩探索和创新，尝试把中国传统转译出当代语言，并以现代的穿着方式演绎传统美。在汲取中国传统工艺的智慧和创造力的同时，希望传达给使用者的是一种时代语境下的、中国式的生活方式。

Xie Yi
Silk and linen

颉逸
丝、麻

Wang Zhihui                                        王志惠

"Saving things for use" is a comprehensive embodiment of ecological aesthetics and values of creation. It pursues the optimal ratio between input and output in creation art on the basis of "harmony", and reflects a sustainable way of life and behavior. This series of design works mainly focus on the redesign of waste fabrics, trying to interpret the concept of sustainable development by recycling waste clothes and giving personal emotion and vitality to the works.

"节物致用"是生态美学、造物价值观的综合体现，其追求的是在"和谐"的基础上造物艺术中投入和产出之间的最优比率，体现的是一种可持续发展的生活方式和行为方式。此系列设计作品主要着重于废旧面料的再设计，试图从废弃衣物回收二次创作来诠释持续发展理念，并为作品赋予个人的情感与生命力。

Symbiotic Series
Native cloth

共生系列
家织土布

Wu Fan / Li Pinyi

吴帆 / 李频一

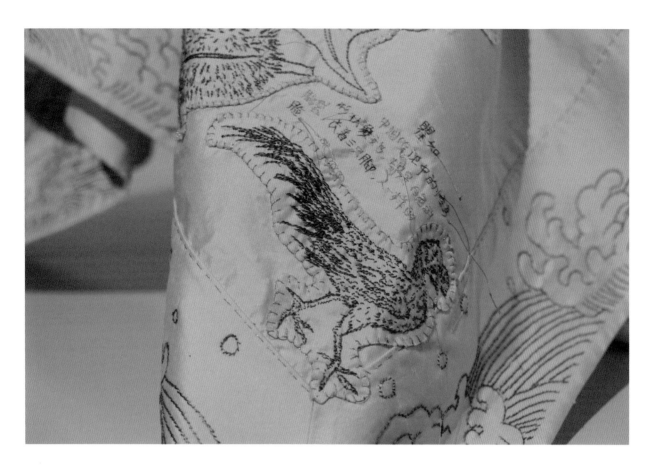

This fashion piece *Ballad for Classic of Mountains and Seas* chooses auspicious animal patterns and text word annotations from *Classic of Mountains and Seas* with Chinese traditional hand embroidery including plain embroidery, patch embroidery, and disordered stitching embroidery. The work vividly represents and explains ancient Chinese romantic feelings, rich imagination and wisdom, and an ideal view of life in a tangible way. The piece adopts a simple and intuitive expressive form, with the unique interaction and interest of fashion art, hoping it will become the "fabric reader" of *Classic of Mountains and Seas* for public communication.

作品《山海谣》选用了《山海经》中一批经典的关于动植物的祥瑞图案及文字注解，用中国传统手工刺绣的平绣、补绣和乱针绣技艺，以可触摸的方式形象地再现和图释中国古人浪漫的山川情怀、丰富的直观想象力和智慧，以及向往美好的原始生命观。作品采用简洁、直观的表达形式，以服饰艺术独有的互动性和趣味性，冀其成为《山海经》的公共传播之"布艺读本"。

Ballad for Classic of Mountains and Seas
Silk

山海谣
丝

Wu Jing

吴晶

The Inspiration of this design comes from the cloud shoulders and the long flowing silk inner sleeves in the opera costume. It adopts the techniques of manual sewing and combines the transformation of old jeans to realize the integration of tradition and modernity and the design concept of green life.

作品灵感来自戏曲服饰中的云肩和水袖，运用手工缝制等，结合废旧牛仔裤的改造，实现传统与现代的融合及绿色生活的设计理念。

Encounter Drama
Cotton

戏缘
棉

Old worn jeans in daily life are selected. The technology of polishing is used to find memory "traces" that the life and time leaves on shabby places of the clothes. *The Vast Land handed down* from several generations is analyzed and the form of "dialysis" security check that people experience in their travelling is used. Besides, "marks" are left to polish rivers and mountains in the structure of dialysis. Besides, the patterns of "traces" are used to express the owner's yearning for beautiful life and the owner's life experience.

选择日常生活中穿旧的牛仔衣裤，并借助打磨的技术在衣物的破旧之处寻找生活与时间给主人留下的回忆"痕迹"。作者借助传世作品《千里江山图》与出行所经历的"透析"安检的形式，并利用留"痕"的手法，打磨出心中的山水与透析的结构，从而通过"迹"的图式，表现主人的美好向往和生活体验。

Mark · Trace
Denim fabric polishing

痕 · 迹
牛仔面料打磨

Dong cloth is a kind of exquisite handicraft textile art unique to the Dong people in southeast Guizhou. It is a symbolized traditional cloth in dong people's clothing. As a symbolic element, it represents the national connotation and regional characteristics of Dong people's clothing culture. The background color of Dong cloth is purple ochre, and its surface is purple, which looks like a unique visual effect of metal plating from a distance. A complete piece of dong cloth is finished after dyeing, hammering, soaking, steaming, drying and other procedures. Its meticulous and complicated production technology can be said to reflect the unique insight and outstanding creativity of dong people. The origin and development, cultural characteristics, material properties, practical functions and production process of Dong cloth are analyzed and studied through field investigation and literature reading in the Dong village of Zhaoxing Township, Liping County, Qiandongnan Autonomous Prefecture, Guizhou Province. This paper explores the possibility of dong cloth in costume design through the techniques of quilting, hollowing, compound, carving, weaving, folding, stacking and corrosion from the aspects of color, form and material.

Wu Qiuyu                                   伍秋裕

侗布是黔东南地区侗族特有的一种手工纺织艺术精品，是侗族服饰中符号化的传统布料，作为标志性元素代表着侗族服饰文化的民族内涵和地域特性。侗布的底色是赭石偏紫，其表面泛紫色光泽，远看如同镀了金属一般的独特视觉效果。一块完整的侗布是经过染、锤、浸、蒸、晾晒等多道工序后才完成，其细致烦琐的制作工艺可以说是侗族人民独特的颖悟力和杰出的创造力的体现。通过对贵州黔东南自治州黎平县肇兴乡侗寨的田野调查和翻阅文献的形式对侗布的起源和发展、文化特征、材料属性、实用功能、制作流程等作分析与研究。深入了解侗布文化，通过色彩、形态、材料这三方面，运用绗缝、镂空、复合、雕花、编织、褶皱、层叠、腐蚀等手法来探讨侗布在服饰设计上的可能性。

Dong Petticoats
To the dong nationality traditional handicraft
art-dong nationality cloth

侗裳
以侗族传统手工艺术品——侗布为主

The Plug & Play collection is inspired by the traditional Chinese wooden craftsmanship of "mortise-and-tenon" joint. The mortise-and-tenon joint has been used by Chinese craftsmen to connect pieces of wooden materials. Although there are many joint types, the basic mortise-and-tenon structure comprises two fundamental components: a "mortise hole" and a "tenon tongue" to fit each other. In every normal case, the tenon either on the end or edge of a piece of wood is plugged to a square or rectangular hole of mortise. The tenon need to be fit the mortise precisely and usually has shoulders that help to fasten, when it fully enters the mortise hole. The mortise-and tenon joint structure is an art of "relationship" that I try to introduce to my fashion concept, in order to build up new relationship amongst all the relative parties in the fashion space.

My application of mortise-and-tenon structure in fashion design aims at bringing the designers, suppliers and wearers into a closer, more interactive and interchangeable personal connection, which comes into shape along with the efforts of recycling and waste control not only in production but also in everyday use.

Bearing the nature of China's "mortise" and "tenon" structure, the disengaging and re-embedding process provides more possibilities of color, form and material in the presentation of clothes without waste; and, with no hesitation, I believe it is an innovative direction to observe and explore the future of sustainable fashion design.

Xiao Yingxian                    萧颖娴

即插即用系列作品的理念源自中国传统木工艺的榫卯结构。榫卯为榫头和卯眼的简称，是接合两个或多个构件的方法，其中构件中的凸出部分称为榫，凹入部分称为卯，主要被用于家具和建筑领域。榫卯结构是一种"关系"的艺术，本作品将该理念引入服装设计中，通过标准化的榫与卯的结构设计，寻找新的服装构成关系。

这种榫卯结构的服装，是服装设计者与穿着者、穿着者与穿着者之间，通过共享关系所带来的可循环利用与减少浪费的产品。

在榫与卯的抽离与再嵌入过程中，服装的构成可呈现更多可能，它意味着不断变化的色彩、廓型、材质，是服装可持续设计的思考与探索。

Plug & Play-Square
Handmade wool felt

即插即用
手工羊毛毡

Xie Mengdi 谢梦荻

Human life relies on the sun, sunlight glitters dust. The arrangement and flow of nanocrystals outside the barrier of clothes, which gives the body a virtual hiding, as well as gives the virtual a real outline.

人类仰赖太阳，日光让尘埃也闪光，纳米晶体的排列和流淌在衣服的屏障外，给身体一个虚拟的隐藏，给虚拟一个真实的联想。

Sunlight—2021
Transparent resin, nylon, nanocoating

日光—2021
透明树脂、尼龙、纳米涂料

Xiong Yi                              熊艺

The series of works *Hills* comes from the cries from the depths of the mountains. The ethnic minority residents in the mountains live by the mountains. Their clothes have been living in the space of the past, the space of the present and the space of the future. In this series of works, the creator renovates the old clothes in a deconstructive way and presents them with a new look, not only to express his fascination with the timeliness and story of clothes, but also to tell the idea that the past and the present can coexist.

作品《山丘》系列来自大山深处的呼喊，山里的少数民族居民，靠山而居。他们的衣服一直存活在过去的空间、现在的空间和未来的空间。本系列作品，创作者用解构的方式，将老衣服翻新，以一种新的面貌呈现，不仅为了表达自己对衣服的时间性和故事性的着迷，更是想诉说过去和当代是可以共存的思想。

Hills
Old homespun cloth

山丘
老布

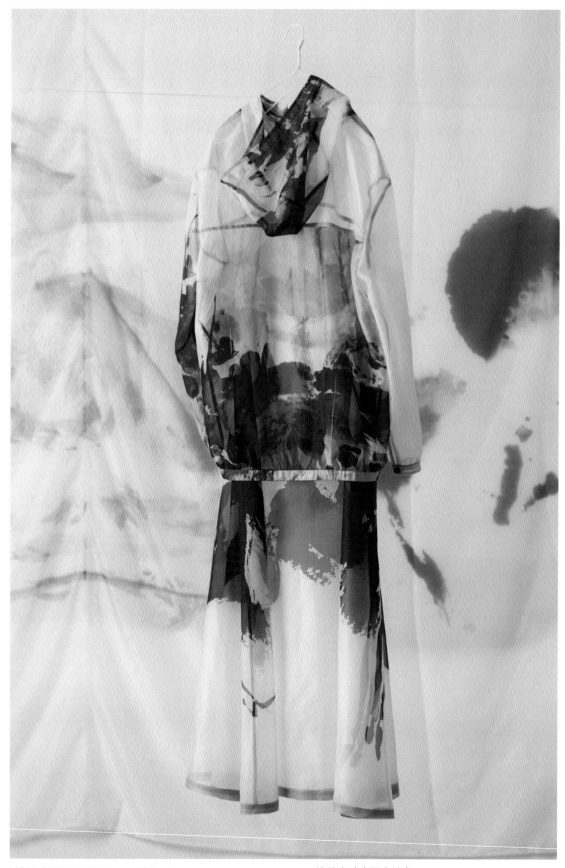

HSU CHIU I. (Taiwan, China)　　　　　　　　　徐秋宜（中国台湾）

Originating from the abstract context of the Orientalist trend, the 2020–2021 art and clothing creations, whether in terms of creative concept, connotation, layout, or creative elements, tend to use the ink imagery of writing pictures and characters, trying to "perceive nature and body The "Dialogue Relationship" is transformed into the Eastern "lyrical abstract" narrative creation.

Through the creation of clothing, patterns and materials have become a new medium of artistic extension, using unfamiliar objects to enter a state of immersion, penetrating directly into the inner micro-fashionable material world, so as to continuously awaken the memory of the simplicity and beauty between the world and the earth. Riding in the virtual boundless universe is a reflection of the assimilation of mind, body and soul with the universe and the integration of art and fashion.

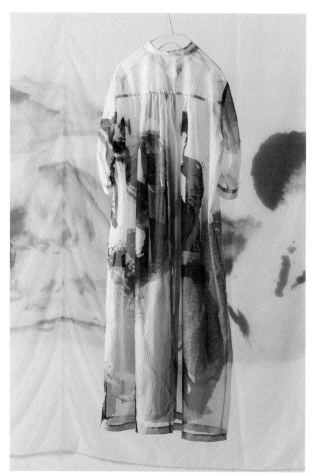

源于东方主义风潮的抽象语境，2020—2021的艺术与服装创作，无论是创作概念、内涵、布局或创作要素，都倾向以写图画字的墨韵意象，尝试将"感悟大自然与身体对话关系"转化为东方"抒情抽象"叙述创作。

透过服装创作，纹样与材质成为艺术延伸的新媒介，用陌生客体进入沉浸状态，穿透直入内层微观时尚的物质世界，藉以持续唤醒天地间劲朴大美的记忆，御风乘虚之际，能无碍地驰骋于虚拟无垠的穹苍邈远之域，是身心灵魂与宇宙太虚同化及艺术时尚合一的映射。

The Body so Alive as Nature.
Satin

时尚川流是动能的整体，一如自然。
丝

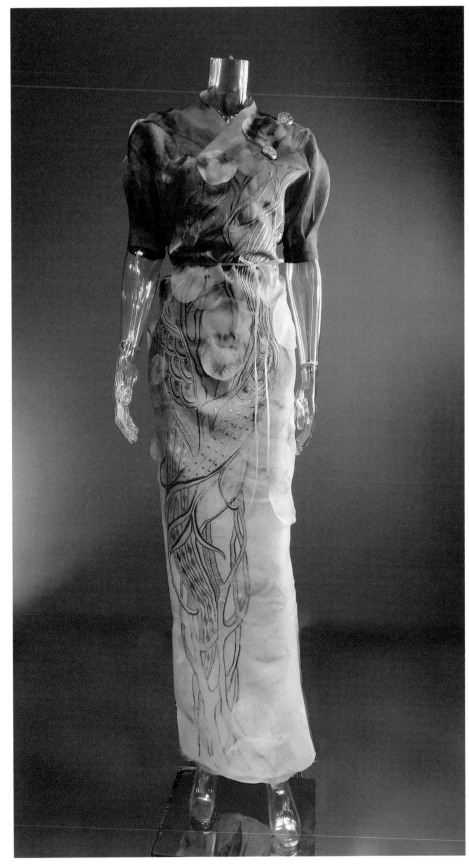

Yan Hongying                                                  闫洪瑛

There is the warm colors and the joy of harvest in Autumn. The light, shadow and emotion of time are presented on silk by plant dyeing and embroidery, as if the space-time shuttle between reality and history, virtual and real changes.

秋季拥有温暖的色彩和丰收的喜悦。植物染色与刺绣呈现出时光影印在绢帛上的情感，仿佛现实与历史之间的时空穿梭、虚实变幻。

Late Autume
Silk organza, embroidery, beading, vegetation dyeing

晚秋
真丝欧根纱、刺绣、珠绣、草木染色

Yan Yishu (Hongkong, China)                                    严宜舒（中国香港）

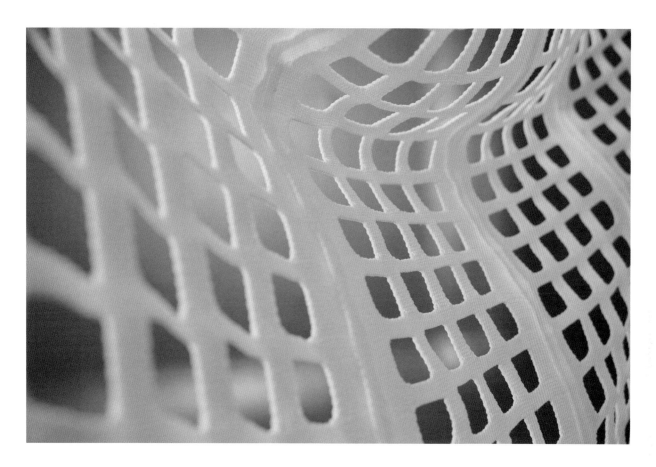

Establishing a matrix that is in the between reality and imagination is always one essential subject in Chinese traditional architecture design. Taking this design philosophy of something from nothing as an inspiration, an exclusive style of open-work fabric has developed through digital knitting technology, in which the hole is specially designed to be square with rounded corners, which metaphors the multiple interpretations of the concept of "Fangyuan" in Chinese in the spatial scale, meanwhile, constructs a plane space with real and virtual. When layering and assembly these 2D cloths to 3D clothing, dynamic and wearable contrast of solidness and void are established with the rise and fall of the body and its movements. Light and shadow, static and dynamic, architecture and fashion, traditional philosophy and modern technology all exist in this "Fangyuan" (dimension).

中国传统建筑设计中，营造虚实相交的环境空间是极为重要的议题。以从无到有的空间构成哲学为灵感，利用数码针织技术设计及编织了特殊的方圆孔洞面料，隐喻文中"方圆"的概念在空间尺度中的多重释义，并同时构建出实中有虚的平面空间。面料层叠交错、组合成为可以穿着的衣衫，平面空间也转向三维立体。随着人体曲线的起伏及身体的活动，一种动态的、可穿着的虚实之间便由此构建出来。光与影、静与动、建筑与服装、传统哲学与现代技术都存在于"方圆之间"。

Fang Yuan
Polyester yarn

方圆
涤纶纱线

David Yeung (Hongkong, China) / Cheung Lio (Hongkong, China) / Au Kim (Hongkong, China)

杨大伟（中国香港）/ 张丽蓉（中国香港）/ 区永欣（中国香港）

SOS
urvive ptimistic taycation

Love under pandemic    Present by David Y & Team

The project is inspired by Covid 19. I called up a team-Kim Au, Lio Cheung and me who work together. We think of SOS based on the issue but the feeling is not danger but SURVIVE, OPTIMISTIC, STAYCATION-We should be positive enough to face it. My fuchsia garment stands for survive because it represents a signal for us to take precaution. Kim's voluminous garment with light colors for optimistic because we need to stay calm and be ready to face difficulties. Staycation means we need to make use of limited space to create comfort and warmth and so Lio's head accessories are created. Lycras and knitted fabrics are keys because of stretch and tenacity properties are good for dance. Bamboos are easy to bend so as to create forms for head pieces. Dancers dress up in full and perform so as to present dynamic and strength against the odds. The whole concept is human should be flexible enough to face changes. We think fashion is not only confined to the area of showing one's identity and beauty in society, but also a way to express what we can see and perceive to echo what is happening in our world. We believe fashion has love and caring and so we use fashion to express the feeling of tension, struggle and release.

S (Survive). O (Optimistic). S (Staycation)
Lycra, knitted cotton for garments, bamboo and stretchy fabrics for head accessories

该项目的灵感来自新型冠状病毒肺炎和口罩。我召集了一个由Kim Au，Lio Cheung和我组成的团队共同致力于这项目。我们基于此来考虑S（生存）．O（乐观）．S（居家度假），但感觉不是危险，而是生存（Survive）、乐观（Optimistic）、居家度假（Staycation），我们应该积极面对这问题。我的紫红色服装代表着生存，因为它代表了我们采取预防措施的信号。Kim 浅色的宽松服装表示乐观，因为我们需要保持镇定并准备面对挫折。小休闲空间意味着我们需要利用有限的空间来创造舒适和温暖，因此 Lio 创造了头饰配件。莱卡纤维和针织物是关键，因为具有弹力和韧性，有利于舞蹈。竹子很容易弯曲，可以制成头饰。舞者盛装表演，表现出活力和力量来克服困难。这意味着人类需要灵活地面对变化。我们认为时尚不仅限于在社会上展现个人身份和美丽的领域，而且是表达我们可以看到和感知的东西的方式，以呼应我们世界上正在发生的事情。我们相信时尚充满爱与关怀，因此我们用时尚来表达张力、挣扎与释放。

S（生存）．O（乐观）．S（居家度假）
莱卡、针织服装、头饰为竹子和弹性面料

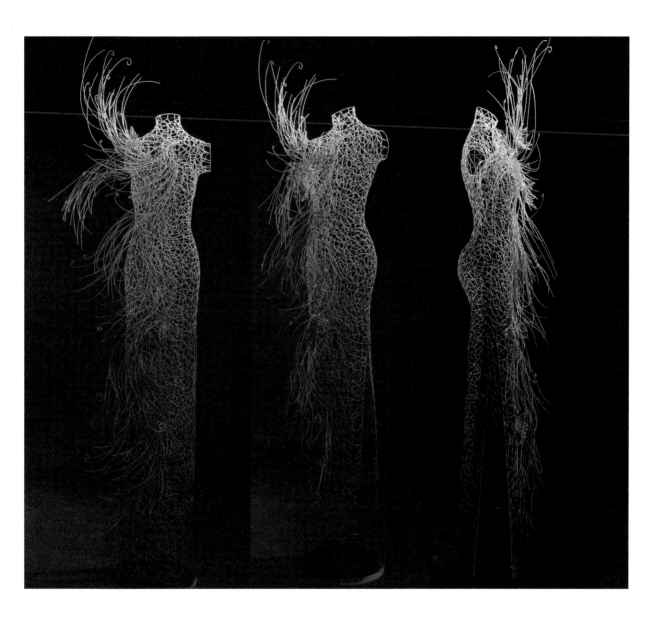

Yang Qiuhua                                         杨秋华

This work selects Chinese traditional cheongsam and Phoenix as the main design elements, adopts modern design techniques, conceptualizes the traditional cheongsam, uses unconventional metal materials, and makes a new interpretation of the cheongsam showing women's softness. The hardness of metal materials forms a strong contrast with the softness of cheongsam itself, but it can be integrated without any sense of violation. Cheongsam itself adopts irregular hollowed out small pieces of different sizes for performance, "phoenix" adopts a smooth linear three-dimensional expression method. The interlacing of lines and surfaces makes the overall shape more flexible and vivid. The hollowed out treatment can produce free visual transformation, so as to produce changeable three-dimensional visual effects for the whole work.

此作品选用了中国传统的旗袍和凤鸟为主要设计元素，采用了现代设计手法，对传统旗袍进行概念化处理，采用打破常规的金属材质，对展现女性柔美的旗袍进行了全新的演绎。金属材质的硬和旗袍本身的柔形成了强烈的对比，但又可以毫无违和感地融为一体。旗袍本身采用了大小不一的不规则镂空小块面进行表现，"凤"则采用了流畅的线型立体表现手法，线与面的交错使整体造型更加灵活生动，镂空的处理则可以产生自由的视觉变换，从而使整个作品产生多变的立体视觉效果。

Phoenix Flying a Century
Iron wire, paint

凤徜百年
铁丝、油漆

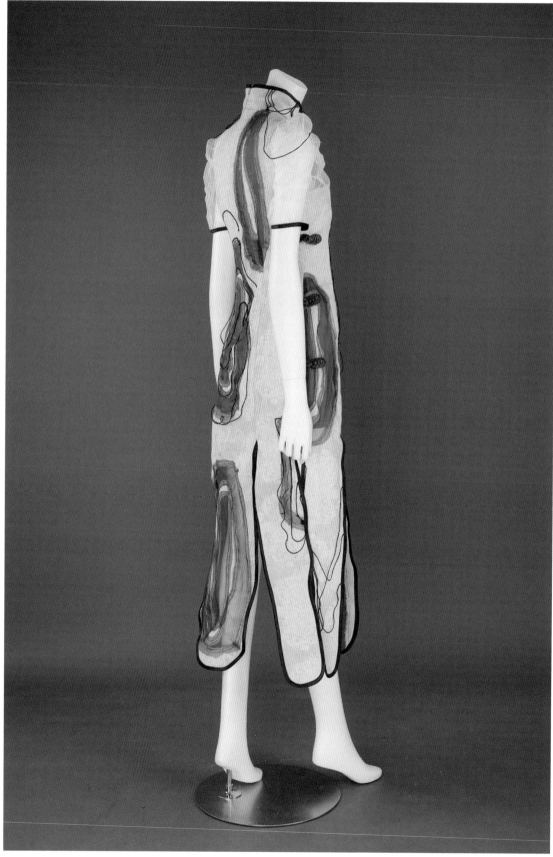

Yu Yiming

于一鸣

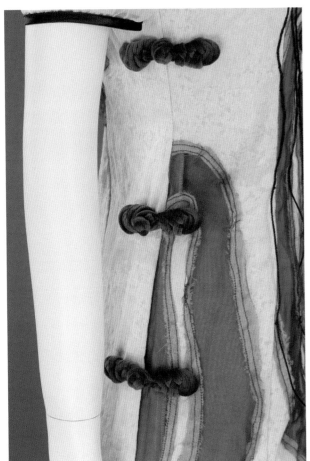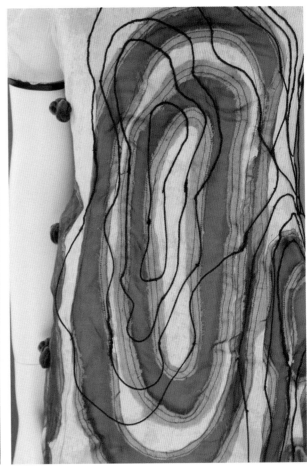

Time never stays for who, but it actually leaves traces of — rings in nature.The rings are like a scale, as if playing the song of the years.Through the hollow layers of fabric stitching, the art reproduces the form of tree rings, but also imitates the reincarnation of time.In the cheongsam with dynamic swing lines, when the dress is driven by wearing, the rhythm of dynamic echo out.

时间从来不会为谁停留，但是它其实在自然界中留下了经过的痕迹——年轮。一圈圈的年轮又如同一个个的音阶，仿佛在演奏着岁月的歌声。通过镂空层次的面料拼接手法艺术再现了树木年轮的形态，同时也模仿着时间的轮回。在旗袍多处加上具动感的回旋线条，当服装被穿着带动时，律动感呼应而出。

Years are Like a Song
Cotton, organza, braided

岁月如歌
棉布、欧根纱、编织绳

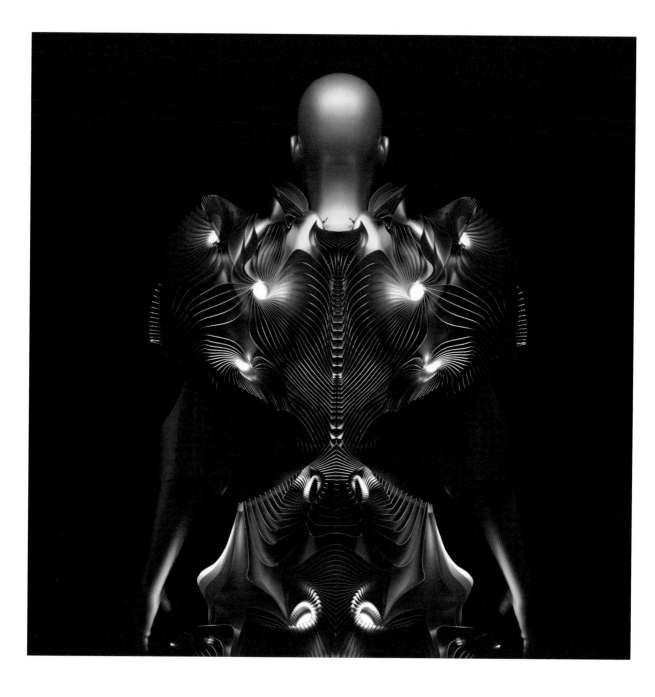

Yu Yimeng / Leon Krykhtin (United Kingdom)　　余一萌 / 利昂·克雷赫廷（英国）

*Dark Matter* is an NFT digital fashion artwork. The universe has its own principle, its own way of forming. The principle is the existence of dark Matter,which act as the invisible force that keeps the universe going.

The work originated from the exploration of the invisible forces of the universe, trying to metaphorically form the similar principles of the universe society and individuals through an artistic way and visualize the invisible laws relationships and emotions. See the unseen, perceive the mystery.

The metaphor principle is expressed in our project by the lines of the Force Fields that drive the pattern logic of the form. The magnetic field force distributes the lines based on bottom-up principle where the machine makes decisions on its form, whereas we create the logic for the machine to design it for us.

作品《暗物质》为一套NFT虚拟时装艺术作品。
宇宙有宇宙的规则，有其自身形成结构的方式，这个规则，就是暗物质的存在，其作为一种无形的力量维持着宇宙的运转。

该作品源起于对宇宙无形力量的探索，试图通过艺术化的方式隐喻形成宇宙、社会和个体的相似原理，将看不见的规律、关系和情感形象化与可视化。感知无形，感知神秘。

在作品创作中，我们通过磁场引力线自下而上驱动生成的方式表达隐喻原则，通过算法创建设计机器来决定最终的分布与形态。

Dark Matter
Iron wire, paint

暗物质
铁丝、油漆

The connection between urban track and nature is blurred, and the a rrangement of neon and light beams is explored in the fog. Each neon light seemed to be a clue, hidden in the mystery of the case. Each light point is the city's detective, blending into the mysterious and lush foggy night.

Yu Chenxi                                       庾晨溪

虚化城市轨迹与自然界的联系，在雾气迷离之中探寻着霓虹与光束的排列。每一盏霓虹仿佛都是一缕线索，隐藏在迷雾案件之中。每一个灯点都是这座城市的侦探，融入神秘又繁茂的雾夜。

How to Solve a Criminal
Horsehair, wool, dupont paper, three-dimensional composite fabric, plastic bag cloth

如何破案
马毛、羊毛呢、杜邦纸、立体复合面料、塑料袋布料

Yuan Dapeng

袁大鵬

The concept of fashion in Chinese classical aesthetics comes from the elephant and takes clothes as the medium. With the physical fashion design as the carrier, we dissolve, disperse, reconstruct, form and color expression and aesthetic space, which are integrated in a needle and a line. The simple and gradual printing and dyeing system has the mechanism of stacking patterns, clear veins, rotation and winding. The seemingly random freehand paintings in the overall pattern inadvertently reflect the rigorous composition and local crisscross, The art of leaving blank complements the undress culture and inherits the classics.

The fabric adopts the traditional natural texture of cotton, hemp and silk, which returns to its origin, just like immersive walking between mountains and rivers. Hand sewn broken thread embroidery, French embroidery, overlapping embroidery, seeded embroidery, pick embroidery and other embroidery and thread embedding design techniques, texture changes show different mechanisms and visual effects. The clothing shape adopts A-type folding and cutting, and the dotted plane echoes with the three-dimensional profile of the sleeve.

中国古典美学时尚概念，由象入境，以衣为媒。由实物时装设计为载体，消解、打散、重构形色的表现和审美空间，融合在一针一线之间，质朴渐变的印染层叠肌理，脉络清晰，旋转缠绕，整体格局中看似随意，却体现严谨构图，局部的纵横交错，留白艺术与宽衣文化相得益彰。

面料采用棉麻、真丝回归本源，恰似身临其境，游走在山水之间。手工缝制的破线绣、法绣、叠绣、打籽绣、挑绣等各种刺绣和埋线设计手法，纹理变化显现出不同的机理视觉效果。服装造型采用A型折叠剪裁，点染层峦叠嶂的平面与袖片立体廓形，遥相呼应。

The Continuity of Clothes: Walking Between Mountains and Rivers
Cotton, hemp, silk

衣脉相承——行走在山水之间
棉、麻、真丝

Yuan Yan 袁燕

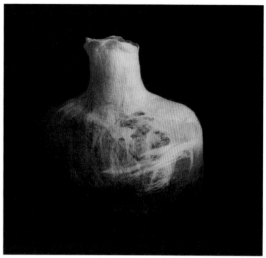

In recent years, I have been studying topics related to the maritime Silk Road. In the research process, I often lament the pioneering spirit of Chinese ancestors to ride the wind and waves. During the field visit, the deep homesickness of overseas Chinese is even more touching. The faded letters and the bright red couplets of overseas Chinese residence all tell the deep feelings.

近些年本人一直研究海上丝绸之路相关课题。在研究过程中常常感叹于中国先民乘风破浪的开拓精神，在实地考察时海外华人深深的思乡之情，更是让人动容。一封封已褪色的书信，海外华人居住地大红色的对联，无不诉说这份深情。

Traces of Memory
Wool, rice paper, wood

迹忆
羊毛、宣纸、木

Yuan Guoxiang (Hongkong, China) / Li Yufan          苑国祥（中国香港）/ 李玉凡

The work is a reflection of the relationship between personal and external space, showing a moment that people involved in the fast pace of life. Nowadays we all live a busy life and enter into each other's vision and space every day, but only left the indistinct image of people and objects passing by quickly. We seem to exist in this spatial dimension as individuals, but do our lives only exist in this dimension?

本作品是对个人空间和外在空间之间关系的思考，展现的是人们在快节奏生活中拥有的一种状态。在这个忙碌喧嚣的城市里，我们每天与无数人进入到彼此的视野和空间里，飞速经快的人们和物体形成一种模糊的图像。可是这不禁让人怀疑，我们看似作为个体存在于这个空间维度，但是我们的生命是否只在这个维度里存在？

Whisking Past
Organza, grenadine

匆匆过去
欧根纱、网纱

Zeng Fengfei

曾凤飞

The source material for this series comes from Guangdong silk specialty Liangshachou, which is also known as "xiangyunsha" and "vinyl silk", which is the collective name of crowfoot and crowfoot. It has the characteristics of soft and smooth texture, strong and wear-resistant, cool on the body, easy to wash and quick-drying. This series of clothing is designed as wedding clothing, using the embroidery craftsmanship of Su embroidery. The women's clothing is the phoenix pattern, and the men's clothing is the Qilin gifting child pattern. Liang sha chou and auspicious patterns are associated with auspicious patterns. While carrying forward the excellent traditional clothing culture, it seeks the meaning of auspiciousness, wealth and prosperity.

该系列制作源材料来自广东丝绸特产莨纱绸，莨纱绸又名"香云纱""黑胶绸"，即莨纱、莨绸的合称。它具有质地软滑、坚挺耐磨、着体透凉、易洗快干的特点。该系列服装设计为婚庆服装，采用苏绣刺绣工艺，女装为凤凰图案，男装为麒麟送子图案。莨纱绸与吉祥寓意的图案结缘，在发扬优秀传统服饰文化的同时，以求吉祥、富贵和兴旺之寓意。

Liangyuan
Liangshachou

莨缘
莨纱绸

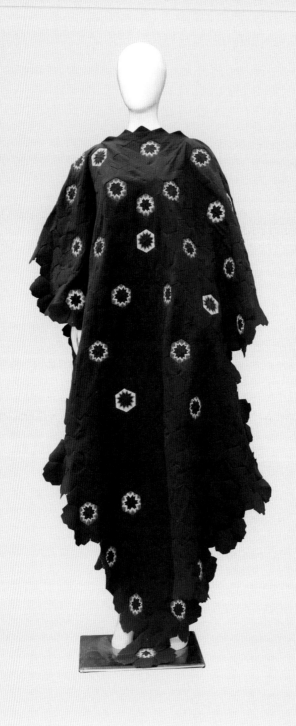

Zhang Gang                                    张刚

Piecemeal fabric, complex manual, piece splicing, is the clear number of pieces, is the idea of cutting constantly! Between the scattering and the reunion, on and off, over and over again, looking forward to some kind of unpredictable show!

零碎的布料、复杂的手工、片片的拼接，是剪不断的思绪！在打散与重聚之间，分分合合，反反复复，期待某种不可预见的呈现！

Cutting Cloth Under the West Window Together
Cotton cloth and fragrant cloud yarn

共剪西窗
棉布 、香云纱

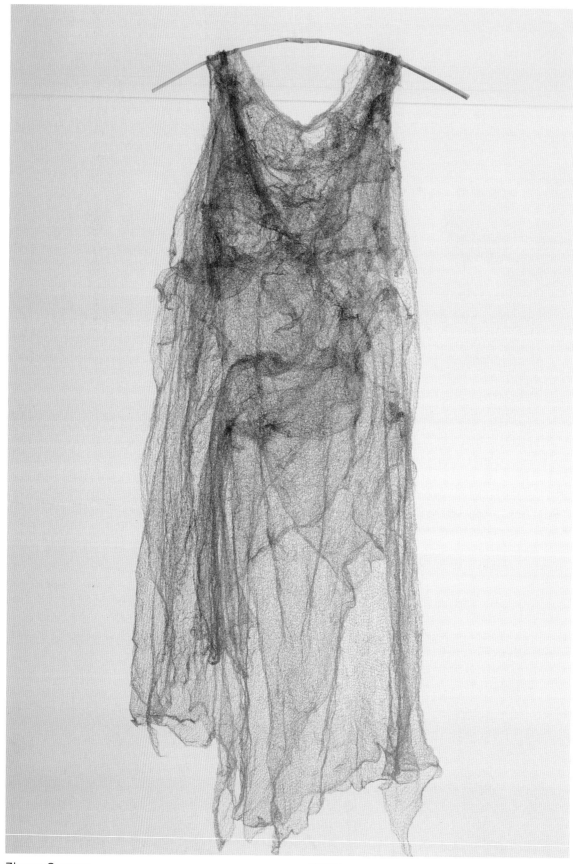

Zhang Guoyun

张国云

This work uses 0.1mm copper wire as thin as hair as material, ancient manual hook weaving as skill, and traditional dress symbol-cheongsam as modeling language. In the process of weaving, the author searched for and collected 0.1mm copper fiber from different regions such as Yibin, Chongqing, Nanning, Southeast Guizhou, Baoshan, Hanoi, (Vietnam) and hue along the Southern Silk Road, and crocheted and combined it to form a suspender skirt with contemporary language and traditional cheongsam dress symbols.

Copper fibers are interspersed between crochets, which makes the smooth copper wire have a sense of texture. It is wrapped in the movement of hand and crochet, extending infinitely, with "infinite" meaning. The hands rub against the copper wire, and the hand temperature, air and copper wire are constantly mixed. For a long time, the copper wire fibers of different regions, nationalities and countries gradually change from golden to black. The passage of time inadvertently reveals in the color transformation of the "line", links the historical origin of the cultural dissemination of the Southern Silk Road in China, and presents the historical sense of youth, years and years, And the infinite extension of the future.

Silk Road and Years
Copper wire

本作品用细如毛发的0.1mm铜丝线为材料、用古老的手工勾织为技艺、以传统服饰符号——旗袍为造型语言。在编织过程中，作者沿着南方丝绸之路沿线的不同城市寻找了搜集了中国的宜宾、重庆、南宁、贵州、保山，以及越南的河内、顺化等不同地区的0.1mm铜丝纤维，将其钩织、拼合而成一件具有当代语言的吊带裙与传统旗袍服饰符号。

铜丝纤维在钩针之间相互穿插，使得顺滑的铜丝线有了肌理感，在手与钩针的运动中缠缠绕绕，无限绵延，具有"无穷"的意蕴。双手与铜丝线相互摩擦，手温、空气与铜丝不停地糅合，久了，不同地域、民族、国别的铜丝纤维逐渐由金黄蜕变成墨黑，时间的流逝不经意间在"线"的色彩变换中流露出来，链接出我国南方丝绸之路文化传播的历史渊源，呈现出青春、岁月、年华的历史厚重感以及未来的无限绵延。

丝路·年华
铜丝

Zhang Peng

张鹏

Different worlds, under the same rain. Regardless of the country, nationality and race, we only make the same voice just for human survival under the epidemic situation. The box cut into the shape of mask symbolizes the difference of region and space.

The sudden outbreak of the epidemic makes us focus on the evironmental and ecological protection again. We express our commitment and action to the sustainable development of fashion through green design of combing the remaining plant dyeing scraps after cutting and wearing old clothes.

不同的世界，下着相同的雨，无论国别、民族、种族，在疫情之下只为人类生存而发出相同的声音，裁剪的方框象征地域之异，空间的差别。

突如其来的疫情，让我们把目光再次投向环境与生态保护，通过将裁剪剩余的植物染色碎料和穿旧服装再利用的绿色设计，表达对时尚可持续发展的承诺和行动。

Share the Same
Trials and Hardships Indigo dyed organza, lace, linen

同云雨
靛蓝染色欧根纱、蕾丝、亚麻布

Zhang Qingxin

张清心

My work is titled *Ji*, this Chinese character has a double meaning. one is an ancient literature style for recording things. the other means memory. Industrially produced clothing as a commodity is no emotional significance. but after being wore, it is given a unique characteristics and soul, it seems like life Itself gets imprinted on the clothing.

作品名为《记》，具有双重含义：记是古代文体，记载事物。另外一个含义便是记忆和印记。将自己的衣服复制，并以文字表述我对它的印象，试图探寻工业化成衣的固有形态在被时代淘汰后与人的依存关系。

Ji
Silk

记
欧根纱

Zhang Tingting

张婷婷

Appreciate the implication and expression of red in Chinese poetry. Using wool and cashmere as the main material, red as the main color, non-rolling felt as the creative technique, fashion art as the medium, to show women's life of perseverance and softness.

体味中华诗词："竹含新粉""映日荷花别样红""千里莺啼绿映红""堕地残红色未蔫""泪脸露桃红色重""狂风落尽深红色""何须浅碧深红色""幂翠凝红色更新""山花红紫树高低""万紫千红总是春""淡淡微红色不深"……感悟古人对"红"的寓意与表达。利用羊毛、羊绒为主要材质，红为主色调，非遗擀毡技艺为创作手法，以时装艺术为媒，展现女性花开花陨飘落红尘的一世美颜……

Woman Flower
Wool, cashmere, silk, cotton thread, passementerie

女人花
羊毛、羊绒、丝、棉线、珠饰

Zheng Chengyuan

郑程元

RED is a dress design based on the expansion of Chinese dress elements such as cheongsam, is a reflection on the future of clothing; The possibility of creating flexible composite fabrics by digital modeling and 3D printing is explored in the new context of clothing. RED is the flow, is the growth, is the trace of iteration, is also the gentle power.

《红装》是以旗袍等中式服装元素拓展进行的穿戴服饰设计，是对未来服饰的思考；以数字化建模，柔性3D打印制作复合面料，探索服饰新语境的可能性。《红装》是流动，是生长，是迭代的痕迹，也是温柔的力量。

Red
3D printing of TPU composite materials

红装
TPU 复合材料 3D 打印

Jung Hyun (Korea)                                        郑贤（韩国）

Transforming from 2D to 3D using laser cutting technology to the pursuit of Zero-Waste fashion.

利用激光切割技术实现从二维到三维的转变，追求零浪费时尚。

Transforming: 2D to 3D
Double-faced PE

转化：从 2D 到 3D
双面聚乙烯

Zheng Mengze                                             郑萌泽

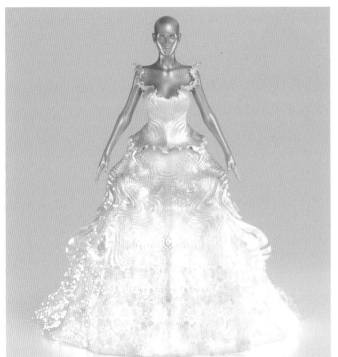 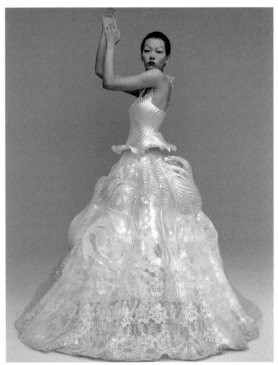

Crystal, a digital wedding dress made of a carved corset with a shell curl, a lace gauze skirt, Parametric gradient crystal skirt and luminous diamond composition. Diamond crystal image. It symbolizes purity, eternity and firmness, and so does the wedding.

这是一件名为Crystal的虚拟婚纱，又可称为水晶婚纱，这套数字婚纱由贝壳卷边雕花胸衣、蕾丝纱裙、参数化渐变水晶裙撑及发光钻石组成。钻石水晶象征着纯洁、永恒、坚固，婚礼亦是如此。

Crystal
Digital virtual

水晶
数字虚拟

Zhou Zhaohui 周朝晖

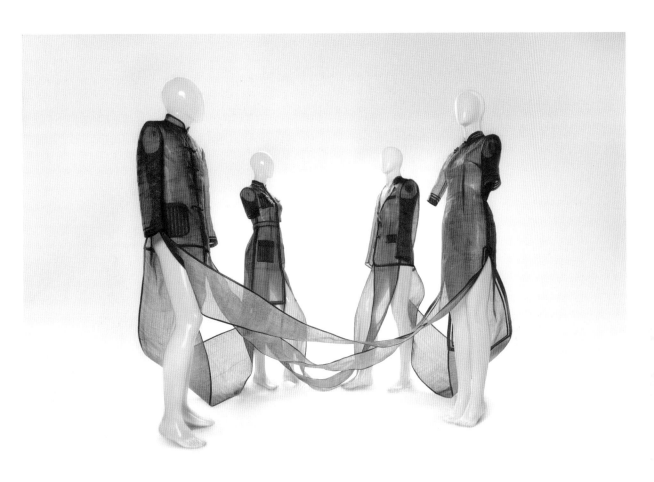

The world under the epidemic is isolated, but Chinese and Western cultures and people are still intertwined and integrated.

疫情之下，世界隔离，但无论中西方文化还是人与人之间，仍是相互缠绕，和合交融。

Entanglements of Destiny
Sinamay

缠
西纳梅麻

Zhou Meng                                  周梦

Interpretation of works:Dong's pleated skirt has a flowing beauty. With its metallic luster, it forms two purple flowing "rivers", which has become the basic tone of the whole dress. It is decorated with silk and silk of the same color system, and then decorated with small embroidered pieces, which completes the transformation between virtual and real, and radiates the new vitality of traditional clothing elements.

侗族百褶裙具有一种流动的美，与它所散发的金属光泽构成两道紫色的流动的"河"，成为整件衣服的基础色调。配以同色系的真丝和绡，再以小绣片点缀其上，完成了虚实之间的转换，焕发出传统服饰元素的崭新生命活力。

Purple Flow
Dong pleated skirt, silk, gauze, satin, embroidered piece

紫流
侗族百褶裙、真丝、绡、色丁、绣片

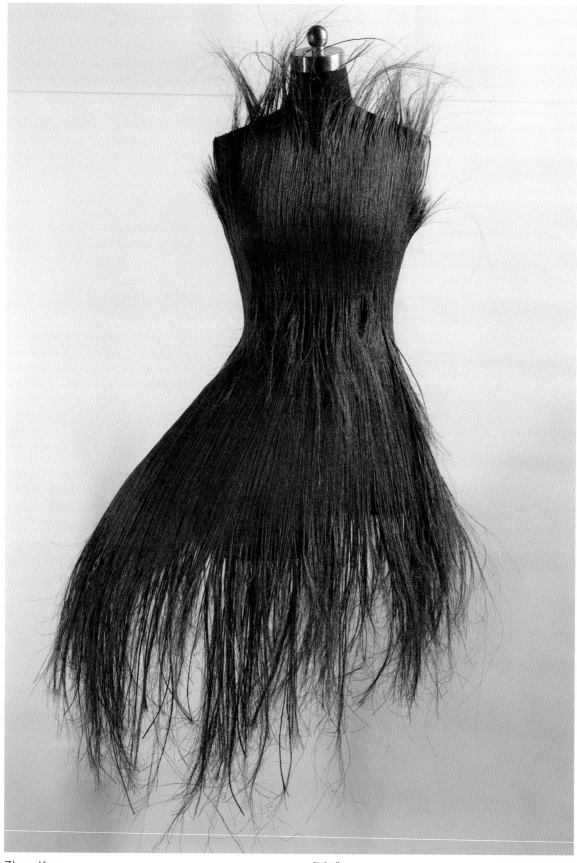

Zhuo Kenan

卓克难

The work takes the Palm Silk form nature as the basic design element and uses the 0.2mm copper wire to gather it into groups, strands, and surfaces. Different with the technique of interweaving longitude and latitude in straw rain cape, this work shows the fit characteristics through the change of density. The silhouette of the garment emphasizes the vertical fall of Palm Silk under the guidance of gravity, which seems like the waterfall. The original natural materials from the forest radiate the fashionable contemporary spirit.

作品采用自然材料棕丝为基本元素，以直径0.2mm的铜线将复数棕丝集合成群、束缚成股、编制成面、围合成体，以贴合身形的方式，塑造出边界上从无到有、从疏到密、复归疏无的有序变化。与传统蓑衣经纬交织的手法不同，作品的廓型更加强调棕丝在重力引导下的竖向垂落，修长成瀑，落于无形。来自森林的原始自然材料，焕发出时尚的当代精神。

Son of Forest
Palm silk, copper wire

森之子
棕丝，铜线

Zong Yuan                    宗源

 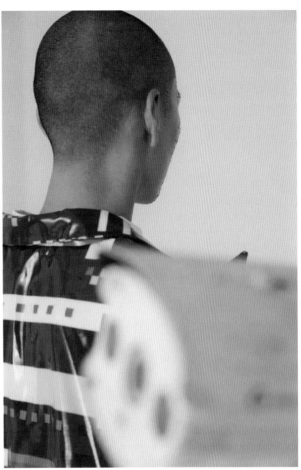

Ubiquitous surveillances, face recognitions, Internet frauds, phone harassments...The rapid development of science and technology may not help to increase the sense of security. In order to resist the intrusion of information, i create some protective means using clothing as a carrier.

When a person is being watched, scanned, or stalked, in order to prevent the other party from obtaining effective information, the outline of the person is blurred by digital means, and the invisibility effect is achieved by combining the way of misunderstanding. Combining glitch art to make the pattern pixelated, combined with the misconception technique, the broken and distorted image expresses a tense and depressive psychological state of urban life.

无处不在的监控、人脸识别、网络诈骗、骚扰电话……科技在不断向前发展的同时,人们的安全感却没有因此增加。我想要制造一些以服装为载体的防护手段来抵抗时代对我们的信息侵入。

当人被注视、扫描、窥探的时候,为了使对方无法获得有效信息,以数字手段模糊人的轮廓,结合视错的方式达到隐身效果。结合故障艺术使图案像素化,结合视错手法,以破碎失真的图像表现了城市生活的一种紧张压抑的心理状态。

Urban Cloaking
Knit fabric

城市隐身
针织面料

Zou Ying

邹莹

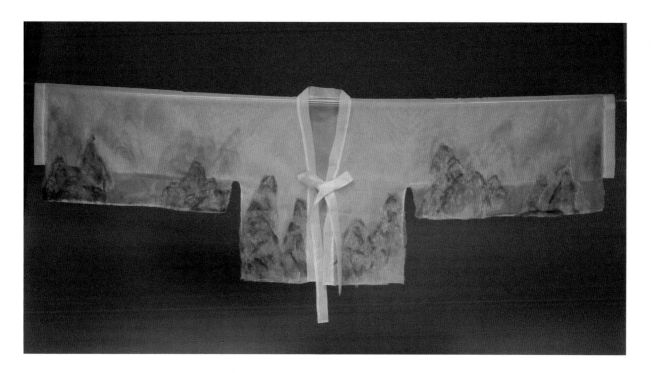

Taking *Qinyuanchun·Snow* in "The Country is So Beautiful" as the theme, taking the green landscape *A Thousand Miles of Country* as the inspiration, with straight and wide sleeves as the shape, pursuing the visual beauty of "layers of mountains and mountains" and "combination of reality and reality" The beauty of the artistic conception creates the style of walking between clothing and wall hangings, three-dimensional and plane, and realizes the style of Jiangshan between light and virtual reality.

Each level of each mountain is individually drawn, cut, burnt, dyed, aligned, and finally fixed with wool felt. Using the light transmission of silk, the change of dyeing, the combination of hand-painted landscape and embroidered landscape, the position and thickness of wool felt, etc., a dozen different levels are created on the seemingly flat clothing. Try traditional clothing, traditional painting and calligraphy. Combination of contemporary spirit and contemporary clothing.

以《沁园春·雪》中"江山如此多娇"为题,以青绿山水《千里江山》为灵感,以直身广袖为形制,追求"层峦叠嶂"的视觉美和"虚实结合"的意境美,创造出在服装和壁挂、立体与平面之间游走的风貌,实现江山在光线与虚实之间呈现的风貌。

每一座山的每一个层次都是单独绘制、裁剪、烧灼、染色、对位,最后以羊毛毡固定的。利用绡的透光、染色的变化、手绘山水和贴绣山水的结合、羊毛毡的位置和薄厚等,在看似平面的服装上打造出十几个不同的层次,尝试传统服装、传统书画和当代精神、当代服装的结合。

Landscape is so Beautiful
Organza

江山如此多娇
欧根纱

# 艺术家简历
# Artists' Resumes

安达
北京

清华大学美术学院硕士毕业；2017年至今，在
北京城市学院艺术设计学部任教。设计作品曾
荣获获丹麦哥本哈根时尚峰会设计二等奖、"纺织
世界杯"第十七届中国（大朗）毛织服装设计大
赛银奖、第三届红衣坊杯世界华服设计大赛获金
奖、中国国际（宁波）年轻设计师大赛铜奖、第
七届中国高等院校设计作品大赛产品设计教师
组二等奖等。参与国家项目《中国工艺美术全
集·国卷·技艺卷》的编撰工作，参与中国流行
色协会科研项目《纺织服装色彩方案预测》编撰
工作等。

安德莉亚·本纳迈德·吉拉利
匈牙利

安德莉亚·本纳迈德·吉拉利是一位优秀的时装
设计师，经验丰富，擅长皮革创作。她是五位入
选2020年纽约独立手提包设计师奖（最佳整体
风格和设计类别）的设计师之一。她也是2018
年意大利国际艺术与设计师大赛的获奖者，2017
年米德兰时装奖决赛选手。

白敬艳
深圳

深圳职业技术学院艺术设计学院服装与服饰设计
专业副教授，教育部职业院校艺术设计专业教学
指导委员会时尚设计专门委员会委员。曾发表多
篇相关服装专业的研究论文及服装设计作品，并
出版过多本服装相关教材，服装设计作品曾入围
第十三届全国美展艺术设计作品展，也曾参加过
学校伦敦设计周展出。

An Da
Beijing

Graduated from the Academy of Arts &
Design , Tsinghua University. Since 2017,
I have been teaching at Beijing City
University. My design works has won
the COPENHAGEN FASHION SUMMIT
Iskool Diploma 2nd. Place,the 17th
CHINA(DALANG) WOOLEN KNITWEAR
Design Competition Silver Award the
3rd HONGYIFANG Chinese Clothing
Design Competition Gold Award,the
CHINA NINGBO INTERNATION YOUTH
COLLEGE STUDENT FASHION Design
Contest Bronze Medal the 7th China
University Design Competition Second
Prize of Teacher Group. Participate in
the compilation of *Chinese Arts And
Crafts·China·Handcraft* (National projects),
*Prediction Of Textil And Garment Color
Scheme* (Research project by China
Fashion & Color Association).

Andrea Benahmed Djilali
Hungary

Andrea Benahmed Dijali is a qualified
fashion designer, highly experienced and
specialised in leather creations.
Andrea is honoured to have been selected
among the 5 finalists of the Independent
Handbag Designer Awards Finalist The
Best Overall Style and Design Category in
New York 2020.
She is also the Award Winner of the
International Art & Designer Competition
2018 in Italy as well as Midland Fashion
Award Finalist 2017.

Bai Jingyan
Shenzhen

Associate professor of fashion and
apparel design, School of Art and
Design, Shenzhen Polytechnic; member
of Fashion Design Committee in the Art
Design Teaching Steering Committee of
Vocational Colleges, Ministry of Education.
Published a number of academic papers,
clothing design works and clothing-related
textbooks. The clothing design works have
been selected for the 13th National Art
and Design Exhibitions, and participated
in the London Design Week Exhibition.

**鲍殊易**
沈阳

鲁迅美术学院染织服装艺术设计系　副教授、研究生导师
中国文物学会纺织文物专业委员会　会员
中国艺术人类学学会　会员
服装设计师协会　会员
与子旗袍服装工作室　设计总监

**鲍怿文**
北京

于2014年就读于中央美术学院设计学院，2015年进入时装设计专业。2021年获得中央美术学院硕士学位。曾获得2021IYDC青年设计师邀请赛银奖，2021届中央美术学院研究生优秀毕业作品奖，Modern Time Fashion show最佳创意奖项，2019第五届大学生人物造型大赛优秀奖，作品曾入选2019第二届中国时装画大展，参与2017年中央美术学院毕业生学位服设计项目，设计2017年央视春晚服装造型，为Rechenberg品牌设计年轻系列。主要作品《身体旅行》《荆棘》、the slience of the sea；dream!dream!dream!；Naught；《肠脑》。主要形式有时装艺术，装置艺术等。

**卡门·里翁**
墨西哥

在墨西哥伊比利亚美洲大学学习平面设计；
在瑞士巴塞尔设计学院学习纺织设计；
1983年开始作为独立设计师进行创作；
1983—1986，墨西哥伊比利亚美洲大学教授纺织品印花工艺及设计；
2008—2010，墨西哥阿纳瓦克北墨西哥大学，教授服装设计；
2008—2010，墨西哥迪塞诺中心，教授服装设计；
2014—2015，英国时装与纺织博物馆，开设无裁剪设计工作坊；
从2003开始，与恰帕斯州本地社团开设工作坊；
2014年获国际时装集团最佳墨西哥设计师奖；
2014年获最具价值墨西哥手工艺及当代设计奖；
2015年获第5届西班牙马德里IE高级和奢侈品行业可持续发展荣誉奖；
2016年获国际绞缬染织研讨会青年设计师"卡门·瑞恩"奖。

**Bao Shuyi**
Shenyang

Associate professor, The Department of Textile & Fashion Design, LUXUN ACADEMY OF FINE ARTS
The Textile Cultural Relics Committee of China Cultural Relics Society, member
Anthropology of art, member
Society of fashion designers, member
YU ZI QI PAO clothing studio, design director

**Bao Yiwen**
Beijing

She studied in the design school of the Central Academy of Fine Arts in 2014 and entered the major of fashion design in 2015. She obtained a master's degree from the Central Academy of Fine Arts in 2021. She has won the silver award of the 2021iydc youth designer Invitational Competition, the 2021 excellent graduate work award of the Central Academy of fine arts, the best creative award of the modern time fashion show, and the excellence award of the Fifth College Student figure competition in 2019. Her works have been selected in the second 2019 Chinese fashion painting exhibition and participated in the graduate dress design project of the Central Academy of Fine Arts in 2017, Design the clothing style of 2017 CCTV Spring Festival Gala and design the young series for Rechenberg brand. Main work, *Body travel, The thorn of the sea, dream! dream!dream!, Naught, Gut brain*. The main forms are fashion art, installation art, etc.

**Carmen Rion**
Mexico

Graphic Designer, Universidad Iberoamericana México,
Master Textile design, Allgemeine Kunsgewerbeschulle, Basel, Switzerland
Independent designer since 1983
Design process for printing textiles Universidad Iberoamericana CDMX, México 1983–1986
Fashion design process, Universidad Anahuac, 2008–2010
Fashion design process, CENTRO de Diseño, Cine y Televsión, 2008–2010
Design Without cutting workshop FTM London, 2014 and West Dean UK, 2015
Workshops with Indigenous groups in Chiapas, since 2003
Fashion group International, "Best mexican designer 2014"
Great Values "Mexican craft and contemporary design 2014"
IE Award in Sustainability in the Premium and luxury sectors, 5th edition 2015 Madrid Spain "Carmen Rion Awards"
for young designers, during Shibori Symposium, Oaxaca, 2016

曹宇培
深圳

本科：北京服装学院2008~2012
硕士：清华大学美术学院2014~2016英国皇家
艺术学院访学
2011年第四届常熟服装城杯中国休闲装设计精
英大奖赛 金奖
2015年第三届西柳杯中国北派服饰设计大赛 金奖
2015年获得美国AOF时装艺术竞赛Maison
Lesage Award 第一名
2018年大浪杯中国女装设计大赛 银奖
2018年12月美国纽约etfashion GDA女装组金奖
2019年作品《解码》入选中国文化部主办的第
十三届全国美术作品展览艺术设计展
2019年作品《对流》入选第二届粤港澳大湾区
学校美术与设计作品展暨第四届广东省高校设计
作品学院奖双年展并且获得三等奖

陈艾
成都

2005年7月至今，就职于四川师范大学服装与设计
艺术学院教师
2011年获四川师范大学美学专业哲学硕士学位
2014年获中央美术学院艺术设计专业艺术硕士
学位。

陈婵娟
美国

美国北德克萨斯州大学助理教授，曾任美国肯特
州立大学助理教授。她的创作目光始终聚焦于可
持续发展服饰设计上，并曾获得多次国际时装比
赛的大奖。她曾连续两年获得美国国际纺织服饰
协会颁发的最佳可持续发展设计奖。她也曾获得
美国俄亥俄州的艺术委员会颁发的个人成就奖。
并多次受邀在加拿大、韩国、美国、中国和英国
等国际服装展发表设计成果。

Cao Yupei
Shenzhen

Ba: fashion design, Beijing Institute of
fashion, 2008 ~ 2012
Ma: fashion design, School of fine arts,
Tsinghua University, 2014 ~ 2016
Royal College of Art
Gold medal of the 4th Changshu clothing
City Cup China casual wear design elite
Grand Prix in 2011
Gold medal of the 3rd Xiliu Cup China
Northern fashion design competition in 2015
In 2015, it won the first place in the
Maison Lesage award of AOF fashion art
competition in the United States
Silver medal of 2018 Dalang Cup China
Women's design competition
Gold medal of women's wear group of etfashion
GDA, New York, USA, December 2018
In 2019, the work decoding was selected
into the 13th National Art Exhibition Art
Design Exhibition hosted by the Ministry
of culture of China
In 2019, the work convection was selected
into the second Guangdong Hongkong
Macao Dawan District School Art and
design works exhibition and the fourth
Guangdong University design works
Academy Award biennial exhibition, and
won the third prize

Chen Ai
Chengdu

From July 2005 till now, I have been
working as a teacher in Fashion and
Design Art College of Sichuan Normal
University.
In 2011, I received my MASTER of
Philosophy degree in aesthetics from
Sichuan Normal University.
In 2014, I received my MASTER of Fine
Arts degree in Art Design from Central
Academy of Fine Arts.

Chen Chanjuan
United States

Chen Chanjuan is an assistant professor at
the University of North Texas in the United
States. She joined the university after
teaching as an assistant professor at Kent
State University, USA. Her research has
focused on reducing textile waste and
increasing garment longevity. Her designs
have been recognized by numerous
international academic and professional
organizations with awards. For two years
in a row, she has received the Educators
for Socially Responsible Apparel
Practices Award for Sustainable Design at
the International Textile and Apparel
Association. She has also awarded by
the Ohio Arts Council with the Individual
Excellence Award in Recognizing
Individual Artistic Achievement. She has
been invited to present and exhibit her
research and design in Canada, United
States, China, South Korea, and the United
Kingdom.

陈华小
广州

现任教广州南洋理工职业学院，2019年作品《丝影·锦色》系列二套被"中国丝绸博物馆"永久收藏；2018年作品《形脉相衣》获"2018中国国际时装创意设计大赛"年度创意奖；2018年作品《锦来衣序》获第十二届中国——东盟青年艺术品创作大赛"一等奖"。

陈炜
桂林

服装与服饰设计专业教师，美国加州州立大学访问学者，苏州大学博士毕业，江苏江南丝绸文化博物馆服饰文化创意总监。多年从事服装艺术设计专业教学，从事服饰心理与服装艺术表达研究，立足地域服饰文化资源，聚焦传统民族服饰艺术的活化与创造性转化，促进区域服饰文化的传承与发展。主持完成江南丝绸文化博物馆创作项目；主持广西区哲学社会科学研究项目，参与多项省部级科研项目，出版相关学术专著，在国家级以上期刊发表论文及专业作品十余篇。

陈闻
大连

中国服装设计协会　副主席；
欧亚（法国）文化艺术交流协会　艺术总监；
东华大学上海国际时尚创意学院　教授；
中国时装设计 金顶奖 设计师；
1996年毕业于中国纺织大学服装学院（现东华大学），获硕士学位，并留校任教。
1997年首次赴法国参加国际博览会并举办首场时装发布会以来，曾于美国、日本、韩国、泰国、埃及等国家举行服装发布、作品展示及主题演讲。
2006年创办上海陈闻服装创作工作室，同时推出"CHENWENstudio"时装品牌。
曾担任过二十余家中外知名品牌的设计总监或品牌顾问。
撰写过多部专业著作和高等服装教育专业教材。
担任多项服装专业赛事评委。

Chen Huaxiao
Guangzhou

Teaching at Guangzhou Nanyang Polytechnic College. Two sets of *Silk shadow· Brocade color* series works were permanently collected by the China Silk Museum in 2019.
The work *Xingmai Xiangyi* won the Annual Creativity Award of 2018 China International Fashion Creative Design Competition, and the work *Jinlai Yixu* won the First Prize in the 12th China ASEAN Youth Art Creation Competition.

Chen Wei
Guilin

A professional teacher of Fashion Design Department of Guangxi Normal University School of Design, a visiting scholar at California State University, a Ph.D. from Soochow University, and creative director of costume culture at Jiangnan Silk Culture Museum. Engaged in the teaching of clothing art design for many years, Engaged in clothing psychology and clothing art expression research, focusing on the activation of traditional ethnic clothing art and creative transformation to promote the inheritance and development of regional clothing culture based on regional clothing cultural resources. Presided the creation project in Jiangnan Silk Culture Museum; presided Guangxi District Philosophy and Social Science Research Project. Participated in a number of provinces and ministries Level scientific research projects, published related academic monographs, published more than ten papers and professional works in journals above the national level.

Vincent Chen
Dalian

Vice president of China Fashion Designers Association
Art Director of Europe-Asia Artistic and Cultural Exchanges
Professor of Shanghai International College of Fashion and Innovation, Donghua University
"Jinding Award" of China Fashion Designer.
In 1996, Chen Wen graduated from China Textile and Clothes College (currently Dong Hua University) with a Master's degree. In 2006, he founded Chen Wen Fashion Design Studio in Shanghai, and launched the brand "CHENWENstudio".
He participated in shows from the first one in 1997 at Bordeaux International Expo to those in US, Japan, Korea, Thailand and Egypt, where he presented his collections and gave lectures. He has released fashion shows featured in business casual men's wear and women's wear and denim.
Chen Wen has worked as art director and consultant for many famous brands.
He wrote several books on fashion illustration and fashion design, including one textbook for university education.
He has been invited as judge for many competitions and contests in the field.

陈燕琳
北京

本科毕业于中央工艺美院服装设计专业，硕士毕业于意大利Marangoni时尚设计学院，博士毕业于中国艺术研究院设计学专业。服装与纤维艺术作品多次参加国内外专业性展览，获得多次奖项。

成昊
北京

成昊，中国服装设计师协会会员，美国ACIC国际认证亚洲服装设计艺术家。他的服装作品多次发布于美国纽约国际时装周、伦敦时装周、中国国际时装周、上海国际时装周、中国青年设计师时装周等。从2017年开始，在服装中加入蜡染技艺和元素。2019年4位银发奶奶身穿他设计的蜡染礼服在三里屯走秀火爆全网。2020年，蜡染系列作品《光阴的故事》首次以中国非遗专场形式亮相伦敦国际时装周。

程琦
重庆

四川美术学院服装系副主任，硕士研究生导师，重庆服装设计师协会理事，重庆市美协设计艺委会委员，中国服装设计师协会成员，英国伯恩茅斯艺术大学访问学者。
主要研究领域：时尚设计、服饰艺术，重点研究传统材料与技艺的传承与创新，传统夏布材料的时尚化运用。
主持传统夏布的时尚研究国家级项目，获国家艺术基金青年艺术创作人才项目。
作品入选2019新西兰世界可穿着大赛，第十一至十三届全国美术作品展，北京国际设计周，第九届全国手工艺展，第十一届从洛桑到北京国际纤维艺术双年展，天然染色展等，致力于传统材料在当代时尚中的传承创新。

Chen Yanlin
Beijing

Bachelor's degree,graduated from the Central Academy of Arts and Crafts with a bachelor's degree in fashion design in 1990 Master of Professional Design in Accessories Design Istituto Marangoni, Italy in 2007 Ph.D. in Design Art, China Academy of Art in 2013.
Clothing and fiber art works have participated in professional exhibitions at Domestic and international for many times,Won multiple awards.

Cheng Hao
Beijing

Cheng Hao, member of China Fashion Designers Association, ACIC International Certified Asian fashion design artist. His clothing has been released many times in the United States New York International Fashion Week, London fashion week, China international fashion week, Shanghai international fashion week, fashion week of young Chinese designers. Starting in 2017, Batik techniques and elements will be incorporated into clothing. In 2019, four silver-haired grandmas wearing his batik dresses went viral on the Sanlitun catwalk. In 2020, the Batik series *The story of Time* made its first appearance at London International Fashion Week in the form of a Chinese non-heritage exhibition.

Cheng Qi
Chongqin

Deputy Director of Department of Fashion, Sichuan Fine Arts Institute; Master Tutor; Director of Chongqing Fashion Designers Association; Member of Design Art Committee of Chongqing Artists Association; Member of China Fashion Designers Association; Visiting Scholar of Arts University Bournemouth, UK.
Main Research Areas: fashion design, fashion art, the inheritance and innovation of traditional materials and techniques in contemporary fashion, as well as trendy traditional grass cloth.
Projects: Head of the national-level research project of the fashionable use of traditional grass cloth; Head of The Young and Creative Art Talent Project of National Art Fund.
Works on Display: 2019 World of Wearable Art Design Competition in New Zealand; The 11th to 13th National Art Exhibition; Beijing International Design Week; The 9th National Handicraft Exhibition; The 11th "From Lausanne to Beijing" International Fiber Art Biennale; The Natural Dye Exhibition.

邓鹤
北京

中央美术学院设计学院时装艺术研究生

杜文
北京

现任北京工业大学艺术设计学院服装设计系教师
2014年受邀参加北京国际设计周"国润华章"的
邀请展
2015年参加"间"——2015中韩Fashion Art交流
展
2015年参加"绽放"——2015时装艺术国际展
2016年参加"蓝之韵"——2016时装艺术国际展
2019年参加千年对话——时装艺术国际特邀展
（中国重庆）
2019年参加深圳国际植物染艺术展
2019年底参加日本东京艺术展
2020年参加新世界·新视界——2020 时装艺术国
际线上展

朴雅茜
西安

1993年5月出生于陕西西安。2017年12月以一
等荣誉学位毕业于全球前十的独立艺术学院——
英国格拉斯哥艺术学院，在校期间攻读服装设计
专业,主要研究方向为刺绣与印花技术在男装流行
服饰中的创造性应用、图案设计和材料创新。学
习期间持续密切关注国内外时尚动态，在读期间
担任助教先后从事了多项与时尚产业密切相关的
工作，从中形成了较为全面的时尚认知，同时积
累了丰富的相关实践经验。现为西安美术学院服
装系专任教师。

Deng He
Beijing

Graduate student in the school of design,
Central Academy of fine arts

Du Wen
Beijing

At Beijing University of Technology
College of Art and Design
In 2014. Invited to participate in Beijing
International Design Week 'National
Runhua Chapter' Invitation Exhibition
In 2015 Invited to participated in 'JIAN'-
2015 China and South Korea Fashion Art
Exhibition
In 2015 Garment Works Participated in
International Fashion Art Exhibition, the
theme of "Blooming"
In 2016 Garment Works Participated in
International Fashion Art Exhibition, the
theme of "Blue Rhyme"
In 2019 Garment works participated in
DIALOGUES 1000 -International Fashion
Art Invitation Exhibition (Chongqing
China)
In 2019 Garment works participated in
"2019 The Third International Natural Dye
Art Exhibition"
In 2019 Garment works participated in
"Tokyo Art Exhibition" in Japan
In 2020 Garments works participated
in 2020 International Fashion Art Online
Exhibition

Du Yaxi
Xi'an

DU YAXI Was born in May 1993 in Xi'an,
Shaanxi Province. In December 2017, she
graduated from Glasgow School of Art,
one of the world's top ten independent art
colleges, with a first-class honours degree.
She majored in Fashion & Textiles, and
her main research direction is the creative
application of embroidery and printing
technology in men's wear, pattern design
and material innovation.
During her study, she continued to pay
close attention to fashion trends at home
and abroad. She also served as a teaching
assistant and engaged in a number
of jobs that are closely related to the
fashion industry, from which she formed
a relatively comprehensive understanding
of fashion and accumulated rich practical
experience. Now she is a full-time teacher
of costume Department of Xi'an Academy
of Fine Arts.

范玉明
苏州

研究员级高级工艺美术师；正高级乡村振兴技艺师；江苏省工艺美术名人；江苏省乡土人才"三带"名人；苏州市工艺美术大师；苏州市第五批非物质文化遗产代表性传承人；缂丝专业委员会副会长。1973年出生于苏州东渚，师从王金山，从事缂丝技艺二十余年，擅长缂人物、花鸟、佛像等。成功复制过"明黄地万历十二章纹福寿衮服""寿桃毡靴""罗地织龙戏珠膝袜"等十余件文物。为提高缂丝技艺水平，进入清华大学工艺美术学院和中国艺术研究院进修学习纤维专业与工艺美术。为传播非遗文化，传授缂丝技艺，与多所学校进行长期合作，教授学生缂丝技艺。

高明
大连

MFE（法国玛洛尼国际时尚教育）市场总监
大连艺术学院　服装学院教师
第三、第四届亚洲设计平面双年展　参展
2018、2020年时装艺术国际展　参展
第十二、第十三届"大浪杯"优秀指导老师
中国当代大学生艺术作品年鉴大赛　评委

高盈
加拿大

蒙特利尔时尚设计师，加拿大魁北克大学教授，HEAD-Genève时尚、珠宝和配饰设计项目前负责人。高盈通过她众多的创意设计项目获得众多个人荣誉，在法国、瑞士、加拿大举办过6次个展，在世界各地参加过100多次群展。创意作品受到了包括《时代周刊》、Vogue杂志、《纽约时报》、Dazed and Confused杂志、《设计时代》ARTE等国际媒体的报道关注。

Fan Yuming
Suzhou

Fan Yuming, researcher level senior craft artist; Senior Rural Revitalization craftsman; Jiangsu arts and crafts celebrities; Jiangsu Province native talent "three belt" celebrity; Suzhou master of Arts and crafts; The fifth batch of representative inheritors of intangible cultural heritage in Suzhou; Vice president of Kesi Professional Committee. Born in Dongzhu, Suzhou in 1973, he studied with Wang Jinshan and has been engaged in the art of Kesi for more than 20 years. He is good at Kesi figures, flowers and birds, Buddha statues, etc. It has successfully copied more than ten cultural relics, such as "Ming and Huang Wanli twelve chapter pattern Fu Shou gun clothes" "Shoutao felt boots" "Luodi weaving dragon playing bead knee socks", etc. In order to improve the skill level of Kesi, he entered the school of Arts and crafts of Tsinghua University and the Chinese Academy of arts to study fiber and arts and crafts. In order to spread the intangible cultural heritage and teach the art of Kesi, we have long-term cooperation with many schools to teach students the art of Kesi.

Arvin Gao
Dalian

Marketing Director of MFE (French Maroni International Fashion Education)
Teacher, School of Fashion, Dalian University of The Arts
The 3rd and 4th Asia Design Graphic Biennale
2018, 2020 International Exhibition of Fashion Art Network
The 12th and 13th "Dalang Cup" Excellent Instructor
Judge of China Contemporary College Students art Works Yearbook Competition

Gao Ying
Canada

Gao Ying is a Montreal based fashion designer and professor at University of Quebec in Montreal, former head of Fashion, jewellery and accessories design Programme at HEAD-Genève, Ying Gao has achieved personal distinction through her numerous creative projects: six solo exhibitions in France, in Switzerland, in Canada, and more than one hundred group exhibitions around the world. Her varied creative work has enjoyed international media coverage: Time, Vogue, The New York Times, Dazed and Confused, Interni, ARTE.

赫然
北京

2017年本科毕业于北京电影学院美术系电影人物造型设计专业。
现研究生就读于中央美术学院时装艺术研究方向。
在多元且丰富的文化与教育背景下，造就了她敏感、柔软的性格和敏锐洞察身边周遭变化的能力。她始终将时装作为媒介和表现手段、尝试不同材质、不同领域与时装结合的可能性。将创作视为精神与情感的寄托，生活中的思考和片段都能捕捉成为其作品的灵感和主题，她试图将作品与自身相结合，找到属于个人DNA的语言。

侯思嘉
南宁

现任广西师范大学设计学院教师，中央美术学院学士，西班牙瓦伦西亚理工大学硕士。

胡园园
广州

就读于广东文艺职业艺术学院。中国图书馆杯主题图像创意设计铜奖获得者。

He Ran
Beijing

In 2017, she graduated from the Department of fine arts of Beijing Film Academy, majoring in film character modeling design.
Now she is a graduate student in the fashion art research direction of the Central Academy of fine arts.
Under the diverse and rich cultural and educational background, she has created a sensitive and soft character and the ability to have a keen insight into the changes around her. She always takes fashion as a medium and means of expression and tries to combine different materials and fields with fashion. She regards creation as the sustenance of spirit and emotion, and the thoughts and fragments in life can be captured as the inspiration and theme of her works. She tries to combine her works with herself and find the language belonging to her personal DNA.

Hou Sijia
Nanning

Teacher of Design of School of Guangxi Normal University; Bechelar of China Central Academy of Fine Arts, Master of Polytechnic University of Valencia in Spain.

Hu Yuanyuan
Guangzhou

Studied in Guangdong Vocational College of literature and art.
Bronze Design Award winner of China Library Cup theme image creativity.

黄刚
深圳

香港理工大学　在读硕士研究生
中国十佳时装设计师
中国针织工业协会　特邀副会长
中国服装设计师协会艺术委员会委员　深圳市时装设计师协会　副会长
"中国工业企业品牌专业人才培养"品牌经理资格证书获得者
广州南洋理工学院客座教授、东莞职业技术学院外聘专业教师。
《天津纺织科技》编委会　专家委员
浙江省毛衫时尚产业创新服务综合体　特约专家导师，河南省淮滨县纺织业发展　高级顾问
设计作品《莞香生产与制作技艺》为东莞市文化馆永久收藏，设计作品
《新长征》主题时装为于都县博物馆永久收藏。荣获第三届中国服饰印花精品大赛"服饰文化奖"，荣获北京2022年冬奥会和冬残奥会制服装备视觉外观设计　优秀设计奖。

黄斯赟
北京

硕士毕业于伦敦时装学院时尚设计与技术（女装）专业，中央美术学院学士。曾多次参与国内外重要赛事获奖和展览，个人作品于国际时装艺术展、中国国际时装周等多个国际知名媒体平台上展示并在全球范围内收获广泛关注。作为任课教师先后为央美时装方向的材料创新、未来时尚与智能可穿戴方向的研究型课题、毕业设计辅导等重要课程主持教学，荣获最佳推荐人奖、指导教师奖。同时，她还拥有丰富的科研和设计实践经验，曾作为团队骨干设计师参与2022北京冬奥会制服与礼仪服装设计、中国人民武装警察部队制服的服装设计、中国花样滑冰国家队运动员服装设计团队设计工作等国家重要科研项目；知名唱作人阿朵、朱婧汐等演艺个人，以及国际著名现代舞团体陶身体作品《10》的合作设计师。致力于智能时装和未来时尚教育研究。

乔安娜·布拉特巴特
法国

设计师坚持最纯粹的巴黎传统进行创作，配饰全部由她在巴黎玛莱区的工作室手工制作。她的反应力和当地文化的认知使她的产品变得独特。她的独家系列自我风格明显，由稀有材料制成，确保客人享有独家定制的、具有法国原创性和技术品质的产品。

Huang Gang

Shenzhen

The Hongkong Polytechnic University
Top ten fashion designers in China
Specially invited vice president of China Knitting Industry Association
Master student in Member of the Art Committee of China Fashion Designers Association the brand manager qualification certificate of "training of brand professionals of Chinese industrial enterprises"
Vice president of Shenzhen Fashion Designers Association.
Visiting professor of Guangzhou Nanyang Institute of technology and external professional teacher of Dongguan vocational and technical college
Expert member of editorial board of *Tianjin Textile Science and technology*
Special expert tutor of Zhejiang sweater fashion industry innovation service complex and senior consultant of textile industry development in Huaibin County, Henan Province;
The design work *Guanxiang production and production skills* is a permanent collection of Dongguan cultural museum. The theme fashion of the design work *new Long March* is a permanent collection of Yudu County Museum.
"Clothing Culture Award" of the third China clothing printing boutique competition Excellent design award for visual appearance design of uniforms and equipment for Beijing 2022 Winter Olympic Games and winter Paralympic Games.

Huang Siyun

Beijing

Graduated from London College of Fashion with a Master's degree in Fashion Design and Technology (Womenswear) and a Bachelor's degree from the Central Academy of Fine Arts. Her work has been showcased at international fashion art exhibitions, China International Fashion Week and many other international media platforms and has gained global influence. As a lecturer, she has been responsible for teaching important courses such as Material Innovation in Fashion, Research-based Projects in Future Fashion and Smart Wearable, and Graduation Design Counselling in Central Academy of Fine Art, winning the Best Recommender Award. At the same time, she also has rich experience in design practice in important national design projects as a team key designer, for the design of uniforms and ceremonial clothing for the 2022 Beijing Winter Olympics, the uniform design for the Chinese People's Armed Police Force, and the athletes of the Chinese figure skating national team; she is the collaborative fashion designer of well-known singer-songwriter A Duo, Zhu Jingxi and other performing individuals, and the international famous modern dance group—Tao 10. She is committed to intelligent fashion and future fashion education research.

Johanna Braitbart

France

It is in the purest Parisian tradition that the Designer Johanna Braitbart creates and manufactures her accessories entirely by hand in Paris in the Marais. Her reactivity and local realisation make her Maison a unique identity. Her exclusive collections, in her own style, are made of rare materials that ensure her customers the originality and quality of French know-how, entirely tailor-made.

凯瑟琳·冯·瑞星博
德国

出生在德国，毕业于巴黎高级时装工会学院。在巴黎接受训练曾为克里斯汀迪奥、香奈儿等品牌工作，在北京工作的高级时装设计师。香云纱（莨纱、莨绸）是她的标志性面料，2000年创立了以自己名字命名的品牌rechenberg，直至今日依然是她设计的主打面料。传统的香云纱自然染色工艺，使用薯莨汁这种含有单宁酸的植物与富含铁元素的河泥作染料，这使她很着迷。受到中国古老奢华的香云纱面料启发，运用香云纱传统的手工染色技术，加上创新独特的理念，巧妙地运用中国的传统中药植物进行面料染色。同时运用了古老的各种绞缬染色法，染出的面料纹理独一无二，反复复杂的渲染工序，自然的植物汁液浸透面料，赋予了面料丰富隽永的色彩。2020年中国纺织非物质文化遗产大会中授予凯瑟琳"传播中国纺织非遗友谊大使"荣誉证书，同时佛山市顺德区伦教香云纱协会授予她"2020年度传播中国香云纱非遗推广大使"荣誉证书。

姜绥祥
中国香港

香港理工大学教授。中国纺织工程学会时装艺术专业委员会副主任。他以金属与纤维为材料从研发物理和化学处理技术着手开展创作研究，艺术和设计在国际上展出，作品被维多利亚和阿尔伯特博物馆、旧金山美术馆和中国丝绸博物馆收藏。

金小尧
北京

"90后"艺术家／教师，毕业于中央美术学院油画系第一工作室，曾获中国绘画新锐奖，全国大学生优秀作品奖，肖像进行时荣誉奖等多个奖项。专注于"基于材料研究的跨界艺术创作与教学"，曾受邀赴马德里美术学院、日本"日下"品牌工厂、荷兰"老荷兰"品牌工厂等地交流考察。2015年至今作为特邀专家在中央美术学院开设绘画材料课程。
曾举办过个人作品展并受邀参加过多个艺术家群展，近年围绕时尚跨界与材料艺术创作了大量作品，并多次入选国内、国际大展。NFT作品在相关平台展览和销售，具有一定的影响力。多件作品被清华大学、中央美术学院收藏，并被编入《中国高等艺术院校教学范本》系列画集。

Kathrin von Rechenberg
Germany

Kathrin von Rechenberg is a German-born, Paris-trained and Beijing-based couture designer. Graduated from Paris's College of Couture "Ecole de la Chambre Syndicale de la Couture Parisienne" she then honed her craft in various Haute Couture houses such as Jacques Fath, Jean Louis Scherrer, Christian Dior, Christian Lacroix and Chanel.
Kathrin von Rechenberg came to China in search of a very special fabric known as "xiang yun sha" or "liang chou" which became her signature fabric when she founded her own label and remains today a cornerstone of her collections. The traditional and complete natural dyeing process of "xiang yun sha" , in which the tannin dye reacts with iron mud fascinates her a lot and inspired her to work with other natural dyes. She creates her own fabrics by experimenting with different traditional and natural dyeing methods, obtaining various structures and an intriguing range of settled colours. in 2020 Kathrin von Rechenberg got awarded the title of "Friendship Ambassador for Textile Intangible Cultural Heritage" by the China National Textile and Apparel Council and recently "Ambassador of Chinese Xiangyunsha Intangible Cultural Heritage" by Foshan Shunde Lunjiao XiangYunsha Association.

Kinor Jiang
Hongkong China

Professor of The Hongkong Polytechnic University. Fashion Art Professional Comimitfee of CTES. He is internationally renowned for his development of the metallized textiles. His art and design projects have been exhibited worldwide. His works have been part of the collections of prestigious Museums including the Victoria and Albert Museum, Fine Arts Museum of San Francisco and China National Silk Museum.

Jin Xiaoyao
Beijing

A post-90s artist/teacher, Graduated from the First studio of the Oil Painting Department of the Central Academy of Fine Arts. He has won the Prize of Chinese painting new talent. National College Students Excellent Works Award, Portrait carried on honorary awards and other awards. For many years, he has been focusing on "interdisciplinary art creation and Teaching based on material Research". He has been invited to Madrid Academy of Fine Arts, Japan "Under the sun" brand factory, Holland "old Holland" brand factory and other places for exchange and investigation.
Since 2015 as the invited experts at the central academy of fine arts courses in painting materials.
He has held solo exhibitions and been invited to participate in several artist group exhibitions. In recent years, I have created a large number of works focusing on fashion crossover and material art, and have been selected for many domestic and international exhibitions. NFT works in the relevant platform exhibition and sales, has a certain influence. Many of his works have been collected by Tsinghua University and Central Academy of Fine Arts, and have been compiled into the series of *Teaching Model of Chinese Art Colleges and Universities*.

兰天
长春

讲师。东北师范大学硕士生，台湾实践大学在读博士，中央美术学院访问学者，绍兴市柯桥区经纬计划A类创业人才。中国国际时装周第十届10+3 SHOWROOM青年设计师基地计划入围设计师。中国服装设计师协会技术委员会委员，国家服装制版师职业资格鉴定考评员，服装制版师二级技师，吉林省服装制版师协会副秘书长。绍兴墨晗文化设计有限公司创始人，MoHanYun品牌主理人。

建筑师和多学科设计师，在伦敦和上海工作。此前，他曾在扎哈哈迪德建筑事务所工作，拥有超过15年的建筑和设计实践经验。他在建筑设计和专业实践方面都有丰富的经验并积极拓展其他领域，许多作品在国际和中国媒体上广泛发表，并在众多国际设计比赛中获奖。

利昂·克雷赫廷
英国

李艾虹
杭州

中国美术学院设计艺术学院副教授，硕士生导师。中国纺织服装教育学会艺术设计教育委员会委员、浙江省美术家协会会员、浙江省服装设计师协会会员、浙江省创意设计协会会员，第三届"中国美术奖·创作奖"国家级银奖获得者，曾多次获得教学成果指导奖国家级荣誉。著有《时装品牌风格设计解码》一书，历年来参加国际时装艺术展览及具有个人代表性的时装艺术作品。

Lan Tian
Changchun

Lecturer. Master student of Northeast Normal University, doctoral candidate of Taiwan Shijian University, visiting scholar of Central Academy of Fine Arts, Class A entrepreneur of Jingwei Plan of Keqiao District, Shaoxing. Shortlisted designer of the 10th 10+3 SHOWROOM Young Designer Base project of China International Fashion Week. Member of technical Committee of China Fashion Designers Association, appellate of national garment plate-making professional qualification examination, secondary technician of garment plate-making division, deputy secretary general of Jilin Province Garment plate-making Division Association. Shaoxing Mohan Culture design Co., LTD. Founder, MoHanYun brand manager.

Leon Krykhtin
United Kingdom

Architect and multidisciplinary designer based in London and Shanghai. Previously he worked at the office of Zaha Hadid Architects and has over 15 years of experience in the practise of architecture and design.
He has extensive experience both in design and professional practice of architecture. Leonid also received prizes and became a finalist in numerous international design competitions. Many of his projects are widely published in international and Chinese media.

Li Aihong
Hangzhou

Associate Professor,Master Tutor,School of Design and Art ,China Academy of Art; Memder of Art Design Education Committee of China Textile and Garment Education Association; Member of Zhejiang Artists Association; Zhejiang Fashion Designers Association; Zhejiang Creative Design Association;winner of thrid National Silver Award of "China Art Award and Creation Award"; won the national honor of teaching achievement guidance award for many times; the author of the book *Decoding Fsahion Brand Style Design*; participated in international fashion art exhibitions and personal representative fashion art works over the years.

李爱舒
深圳

2014年毕业于香港理工大学纺织及服装学系。深圳绒臻至美时尚科技有限公司设计总监。长期从事面料研究和服装设计。

李频一
深圳

从事服装高级定制40年，擅长服装手工制作。曾参与服装作品《棱镜》《结晶体》及纤维艺术作品《方寸论》的制作。

李莲姬
韩国

韩国汉阳大学纺织与服装系教授，在梨花女子大学获得学士及硕士学位。先后于法国巴黎ESMOD时装设计学院及日本时尚针织设计学校学习。

Li Aishu
Shenzhen

 Postgraduate from Institute of Textiles and Clothing in The Hongkong Polytechnic University.  Design Director of Shenzhen Rongzhenzhimei Fashion Technology Co., LTD. She has engaged in textile research and fashion design for a long time.

Li Pinyi
Shenzhen

She engaged in couture and handicraft for 40 years. She is good at clothing handicraft. Besides, she has participated in the production of design works such as *Prism*, *Square* and *Circle*, *Square Theory* and etc.

Lee Yeonhee
Korea

Professor, Dept. of Textiles & clothing, Hanyang University,
Seoul, Korea
Graduuted from Ewha Womans University, Seoul, Korea (B.F.A / M.F.A)
Studied in ESMOD PARIS, Stylism, Modelism, France
Japan Vogue Knit Design school, Japan

李洋
北京

清华大学美术学院染织与服装艺术设计系在读硕士，本科就读于上海戏剧学院戏剧影视美术设计专业。研究方向数字服装设计，参数化服装设计，数字设计与数字制造下的服装设计等。作品 Defense 曾获中意青年未来时尚设计大赛创新设计师奖。曾在加拿大 Cirque du Soleil 担任服装制版助理，在英国设计师品牌 WANHUNG 兼任设计与品牌助理。2019 年出任德国著名导演伊万·潘特列夫《沃伊采克》服装设计主创，2018 年任著名导演周可《乱打莎士比亚》服装设计主创。

李薇
北京

清华大学美术学院染织服装系教授、博士生导师；巴黎高等装饰艺术学院访问学者；中华服饰文化研究会副会长、中国传统文化促进会服饰委员会主任委员、中国民族服饰协会理事、中国流行色拼布委员会主任委员、中国纺织工程学会时装艺术专业委员会副主任。2015 年度国家社会科学基金项目评审委员。
中国国际十佳时装设计师荣获 2014 年亚太经合组织（APEC）会议领导人服装做出突出贡献 - 荣誉奖
2009 年《清、远、静》获"从洛桑到北京"第六届国际纤维艺术双年展金奖。
2004 年《夜与昼》获"第十届全国美术作品展"金奖；
个展：2017 年《李薇高级定制服装秀》新疆国际会展中心，《李薇艺术作品展》香港理工大学；
2016 年《李薇艺术作品展》英国皇家艺术学院，《李薇高级定制服装秀》大连 Z28。

李怡文
肇庆

广东理工学院艺术系教师，设计学硕士。研究方向为民族传统服饰与服装研究。发表专业论文 7 篇；参与的项目《壮锦面料在现代服饰设计中研究与呈现》被评为"优秀研究生创新项目"；个人作品《叠续》《穿裙子的蒙太奇》《重生》《染》《蓝色节奏》《壮纹尚音》《斑岩》《蜡染》等作品获得了参展和获奖证书，其中《叠续》作品获得了第十三届创意中国设计大赛一等奖，《穿裙子的蒙太奇》获得了第三届"包豪斯奖"国际设计大赛金奖。

Li Yang
Beijing

Li Yang is a postgraduate student majoring in Fashion Design in School of Fine Arts, Tsinghua University. She received a bachelor's degree in Drama, Film and Television Art Design from Shanghai Theatre Academy. Her research interests include digital fashion design, parametric fashion design, fashion design under digital design and digital manufacturing, etc. Her work Defense won the Innovation Designer Award of Zhongyi Youth Future Fashion Design Competition. She worked as intern costume pattern cutter at Cirque du Soleil in Canada and design and brand Assistant at British designer brand WANHUNG. In 2019, she acted as the chief costume designer in The famous German director Ivan Pantelev's Wojtsek. In 2018, she acted as the chief costume designer in the famous director Zhou Ke's Beating Shakespeare.

Li Wei
Beijing

Professor of Fashion Design Department of Arts and Design Academy, Tsinghua University. PhD supervisor and visiting scholar in École Nationale Supérieure des Arts Décoratifs (ENSAD) in France. Vice-chairman of Chinese Traditional Costume Research Association, Chairman of Committee Member of China Traditional Culture Promotion Council, Member of Chinese Ethnic Costume association, Chairman of Committee Member of China Fashion Color Association (patchwork department), Vice. director,Fashion Art Professional Committee of CTES member of Evaluation Committee of Annual National Social Science Fund Program.
Top 10 Designers of the 20th China International Fashion Week; Honor Award of APEC Leader Conference costume design in 2014.
Pure, Distant, Still-Golden Award of the 6th "From Lausanne to Beijing" international Fiber Art Biennale in 2009.
Night and Day-Golden Award of the 10th National Art Exhibition Award in 2004.
2017: "LiWei Haute Couture Fashion Show" at Xinjiang International Conference and Exhibition Center
2017: "LiWei Arts Exhibition" at The Hongkong Polytechnic University
2016: "LiWei Arts Exhibition" at Royal College of Art in London, England
"LiWei Haute Couture Fashion Show" in DaLian

Li Yiwen
Zhaoqing

The teacher of Department of Art, Guangdong Polytechnic College, master of design. Research direction: the research on national traditional clothing and clothing. 7 professional papers were published, and the project of Research and Presentation of Zhuang Brocade Fabric in Modern Costume Design was rated as 'Excellent Graduate Student Innovation Project'. Personal works such as Overlapping, Montage in Dress, Rebirth, Dye, Blue Rhythm, Zhuangwen Shangyin, Porphyry, Wax Dye etc. were exhibited and the award-winning certificate, in which the work of Overlapping won The first prize of the Italian-China Design Contest of the thirteenth creation and the work of Montage in Dress won the gold medal of The third "Bauhaus Prize" international design competition.

李迎军
北京

艺术学博士，清华美院副教授、博士生导师，中国流行色协会理事，时装艺术国际同盟常务理事委员，中国敦煌吐鲁番学会会员，敦煌服饰文化研究暨创新设计中心研究员，中国服装设计师协会会员，法国高级时装协会学校访问学者。
致力于民族文化与时尚流行的研究，在北京时装周、中国非物质文化遗产服饰秀、江南国际时装周发布系列作品，参加北京国际设计周、艺术与科学国际作品展等数十项专业展览，设计作品荣获"兄弟杯"国际青年设计师作品大赛金奖、法国国际青年服装设计师作品大赛国家奖等多项国际、全国权威专业竞赛奖励。

李玉凡
上海

本科毕业于东华大学服装设计专业，现在读东华大学服装设计研究生。本科毕业设计作品曾入选iD国际新兴设计师奖决赛，并且入选于上海举办的2020年蕾虎呦桃毕业秀。

梁楚茵
中国香港

毕业于伦敦大学，香港大学，香港理工大学2001年全职投入艺术创作，中国中央美术学院客席讲师，中国服装设计师协会讲师，美国Penland School of Crafts可持续性面料－羊毛毡课程讲师，国际绞结染织网络会员；多次为DeTour，Southern Hemisphere Felting Convergence，The Woolmark Company，Fashion Summit举办讲座及工作坊。澳洲纤维艺术会导师，创办"流动自然后染缸"行动，香港理工大学、香港知尊设计学院及香港高等教育科技学院客席讲师及艺术家合作伙伴，创立香港研艺中心、Handmade by Debbie Leung及长衫妞妞品牌，为香港舞蹈团设计和制作服装；近年投入研究香港长衫文化及融入羊毛毡的技巧制作；作品在国内外展览被收藏。

Li Yingjun
Beijing

Doctor of Arts, Associate Professor and Doctoral Supervisor of the Academy of Arts & Design of Tsinghua University, director of China Fashion Color Association, member of International Union of Fashion Arts, member of China Fashion Designers Association, and visiting scholar of France Haute Couture Association School.

Be committed to the research of "Nationa Culture and Fashion" and won a lot of awards in the Fashion Design contests of international and national. Design works "The Hero", "Line Map", "JingWuMen" won the Golden prize of "Brother Cup" the International Youth Fashion Designer Contest, National Award of Paris Youth Fashion Designer Contest in France, innovation technology award of Hongkong "New Series" fashion competition.

Li Yufan
Shanghai

Graduated from Donghua University majoring in Fashion Design, and now she is a graduate student in Fashion Design of Donghua University. her graduation design work has been selected for iD International Emerging Designer Awards finalist, and selected in the 2020 Graduation Show held by labelhood youtopia in Shanghai.

Debbie Leung
Hongkong, China

Graduated from University of London, University of Hongkong, The Hongkong Polytechnic, in 2001 she turned full time artist.
Guest lecturer of Central Academy of Fine Arts, China, Guest lecturer of China Fashion Designers' Association, Penland School of Crafts, U.S. Sustainable fibre – wool felt course. Member of World Shibori Network. Seminars and workshops for DeTour, Southern Hemisphere Felting Convergence, The Woolmark Company, Fashion Summit
Workshop for Fibre Art Australia Founded "Mobile Dye Pot" movement Guest Lecturer of Hongkong Polytechnic University, Hongkong Design Institute and Technological and Higher Education Institute of Hongkong and collaborative artist, Founded Hongkong Craft Center, Handmade by Debbie Leung and Little Ones Adventure, Design and making costumes for Hongkong Dance Company.In recent years research and study the origin and development of Hongkong cheongsam and integrate traditional feltmaking techniques in creating a new generation of cheongsam, Her artwork was exhibited and collected around the world.

梁莉
南宁

现任广西师范大学设计学院教师，广西大学工学学士，中央美术学院文学硕士。曾参加2007、2008、2010、2013、2015、2019年时装艺术国际展。

梁明玉
重庆

艺术家、服装设计师、西南大学纺织服装学院教授、重庆市服装设计师协会主席、时装艺术国际同盟常务理事、中国纺织工程学会时装艺术专业委员会副主任、重庆非物质文化遗产传统技艺《梁明玉缝纫造型》负责人；
2018"时尚北京"时尚艺术终身成就将获得者；
2019年大型装置艺术作品《马塞马拉》受邀在英国伦敦TATE MODERN交互空间展览；
2018年大型装置艺术作品《马塞马拉》，呼吁人类对环境和野生动物的保护。同年，作为主要展出人之一，携作品参加2018《时尚北京·中国时尚4x30——北王南张东吴西梁展》；
2008年担任第29届奥林匹克运动会（北京）开闭幕式服装主创设计师，被奥组委授予"优秀个人"荣誉；
2005年受邀担任第五届亚洲太平洋城市市长峰会（2005 Asia Pacific Cieies Sammie）贵宾服装首席设计师；
曾担任多部获得中国文化部大奖的舞剧服装设计师；
曾担任多次国家运动会开幕式服装总设计师；
曾出版《裳》《创意服装设计学》《从草图到成衣》《服装设计创意空间》。

梁之茵
南宁

清华大学美术学院在读博士。毕业于清华大学美术学院，获文学学士和艺术学硕士学位。曾赴英国皇家艺术学院和帝国理工学院交流学习。
"学院联合杯"全国家居产品创意设计大赛特等奖
"鲁绣杯"第四届全国大学生家纺创意设计大赛金奖
"第十四届全国纺织品设计大赛暨国际理论研讨会"银奖
"第十六届全国纺织品设计大赛暨国际理论研讨会"铜奖
"汉帛奖"第27届中国国际青年设计师时装作品大赛优秀奖
参与展览联合国"无界"国际艺术展
德国"点、线、面"德中艺术与设计主题联展
中国纤维艺术世界巡展——意大利展
亚洲纤维艺术双年展
首届"纺织之光"中国纺织艺术展
国际防染艺术展暨国际理论研讨会
"艺舞春风"丙申艺术名家邀请展

Liang Li
Nanning

Teacher of Guangxi Normal University in Design of School; BE of Guangxi University, MA of China Central Academy. participated Fashion Art International Exhibition in 2007, 2008, 2010, 2013, 2015, 2019.

Liang Mingyu
Chongqlng

Artist, Fashion designer, Professor, Textile College, Southwest University
Chairman of Chongqing Fashion Designers Association
Executive Director of the International Federation of Fashion Arts
Vice director of Fashion Art Professional Committee, CTES.
Responsible person of Chongqing intangible cultural heritage traditional skills Liang Mingyu sewing modeling
2019's large-scale installation *Masai Mara* has been invited to exhibit in the TATE MODERN in London, England.
In 2018, *Masai Mara*, a large-scale installation art work, as one of the main exhibitors, he participated in the 2018 Fashion Beijing China Fashion 4x30-Beiwang South Zhangdong Wuxi Liang Exhibition and was awarded the Fashion Beijing Lifelong Achievement Award for Fashion Art.
Creation of Clothing for the Opening and Closing Ceremony of the 29th Olympic Games (Beijing) in 2008
Designer, who was judged as an outstanding individual by the Olympic Organizing Committee.
In 2005, he was invited to serve as the chief designer of the fifth Asia Pacific Cieies Sammie Summit. He was awarded by Chongqing Municipal People's Government as an advanced individual in the preparation of the 2005 Asia Pacific Mayors Summit.
He has worked as a fashion designer of many dance dramas which have won the prize of the Ministry of Culture of China, and as the chief fashion designer and creative director of the opening ceremonies of many national games.
She has published books as *Clothes*, *Creative Fashion Design*, *China Textile Publishing House*, *Creative Space of Fashion Design*.

Liang Zhiyin
Nanning

PhD candidate of Tsinghua University. BA and MA degree of Tsinghua University.
Exchange student of Royal College of Art and Imperial College London
2019 Excellence Award of Hempel Award the 24th China International Young Fashion Designers Contest
2016 Bronze Award of National Textile Design Contest
2014 Top Award of National Home Products Design Contest
2014 Gold Award of National Textile Design Contest of College Students
2014 Silver Award of National Textile Design Contest
2019 Artwork exhibits in 11th Asia Fiber Art Exhibition and Symposium
2018 Artwork exhibit in Chinese Textile Art Exhibition
2018 Artwork exhibits in The Contemporary Chinese Fiber Art Exhibition in Italy
2017 Artwork exhibits in International Art Exhibition (Youth Art Award)
2017 Artwork exhibit in International Defending Dyeing Art Exhibition
2016 Artwork exhibit in Art Masters Invitational Exhibition
2016 Artwork exhibits in Arts & Design Exhibition of Germany and China

刘锦玉
江门

江门职业技术学院艺术系副教授，广东省美术家
协会会员，承担服装设计、珠宝首饰设计课程教
学。主持完成省级项目多项，在《丝绸》等核刊
及普刊公开发表学术论文、作品多篇，作品多次
参加国家级、省级艺术设计展赛并获奖。

刘瑾
南昌

毕业于苏州丝绸工学院服装设计专业，江西师
范大学美术学院服装与服饰设计系系主任，副
教授，研究生导师，江西省非物质文化遗产赣绣
（豫章绣）非遗研究基地负责人，中国服装设计
师协会会员，江西省美术家协会会员。
作品参展《2021中国国际纤维艺术展》《记
忆·状态》中日现代艺术展、东方印迹—江西师
范大学美术学院与埼玉大学教育学院第一回艺术
作品交流展、"秋实穰穰——庆祝改革开放四十
周年"江西师范大学、南京晓庄学院美术作品
展、江西省首届艺术设计双年展等多个展览。

刘静
重庆

2013年于深圳新百丽鞋业有限公司时尚研究中
心担任产品设计师；
2014年创立个人同名独立设计师品牌LIU JING；
2017年硕士毕业于四川美术学院；
2017年于重庆国际时装周举办个人服装作品专
场发布会；
2019年赴伦敦参加伦敦大学生时装周；
2020年于香港理工大学攻读博士学位；
曾先后获得第十一届"迪尚杯"中国时装设计
大赛金奖、第22届中国时装设计十佳新人奖、
2016年大连杯银奖、第3届华人杯铜奖、第十一
届乔丹杯优秀奖等多个奖项；重庆十佳服装设计
师，曼宝洛羊绒服饰有限公司设计总监，并同时
任教于重庆第二师范学院服装与服饰专业。

Liu Jinyu
Jiang men

Associate professor of Art Department
of Jiangmen vocational and technical
college, member of Guangdong Artists
Association, responsible for the teaching
of fashion design and jewelry design.
She has presided over and completed a
number of provincial projects, published
many academic papers and works in
nuclear and general journals such as
*Silk*, and her works have participated in
national and provincial art and design
exhibitions and won awards for many
times.

Liu Jin
Nanchang

Graduated from Suzhou University fashion
design major, jiangxi Normal University
fashion and costume design department
director, associate professor, postgraduate
tutor, jiangxi intangible Cultural heritage
Jiangxi Embroidery (Yuzhang embroidery)
intangible Heritage research base, member
of China Fashion Designers Association,
member of Jiangxi Artists Association.
Works *2021 China international fiber art
exhibition*, *The state of memory*, sino-
japanese modern art, east, imprinting-
jiangxi normal university academy of
fine arts and the first time the saitama
university education college art
JiaoLiuZhan, "achievements RangRang-
celebrating the 40th anniversary of reform
and opening-up" jiangxi normal university,
nanjing xiaozhuang college of fine arts
exhibition, in jiangxi province, the first
biennale art design And so on.

Liu Jing
Chongqing

Worked as a design assistant in Belle
International Holdings Limited in Shenzhen in
2013; Created brand "LIU JING" in 2014; MA in
Sichuan Fine Arts Institute in 2017; Hold an
personal clothing works show in
Chongqing International Fashion Week in 2017;
Parted in London Graduate Fashion
Week in 2019; PhD in The Hongkong
Polytechnic University.
Got the Gold Medal of the 11th China Fashion
Design Contest Award, the Top 10 of
the 22nd China Fashion Young Talent Award,
Silver Medal of 2016 Dalian Cup International
Youth Fashion Design Contest
Bronze Medal of the Chinese Fashion Design
Contest Excellence
Award of 11th Qiao Dan Cup China Sports Wear
Design Contest and etc.
Top 10 Fashion Designer of Chongqing
Design director in MARBALUS Clothing
Co., Ltd , the lecturer of Fashion Design
Department in Chongqing
University of Education.

刘君
深圳

教授
深圳职业技术学院文化创意产品研究院副院长
苏州大学艺术学院学士
武汉理工大学设计学院硕士
英国南安普顿大学温切斯特艺术学院访问学者
中国丝绸博物馆时尚专业咨询委员会专家委员
中国服装设计师协会理事
时装艺术国际同盟常务理事
深圳市服装设计协会常务副主席
深圳市宣传文化事业发展专项基金评审专家
深圳市政府创意设计七彩奖评委
联合国教科文组织创意城市网络深圳创意设计新
锐奖评委
中国美术家协会会员


刘沛霖
深圳

毕业于深圳职业技术学院艺术设计专业
2020年国际大学生艺术年度奖 优秀奖
2021年获得新加坡金沙艺术节 双铜奖
2021年作品《START x START》《星际探索》
入选第四届粤港澳大湾区高校美术与设计作品展
暨第六届广东省高校设计作品学院奖双年展
2021年获得第三届台湾国际大学生年度艺术奖
铜奖和优秀奖


刘娜
天津

天津美术学院副教授，天津市美术家协会会员多
件作品入选国际、国内专业性展览，在国际纤维
艺术双年展中曾获得铜奖1次、优秀奖数次，且
受邀参加中国当代纤维艺术世界巡展，曾参加在
巴西、波兰、哥伦比亚、厄瓜多尔、韩国等国家
的巡展。多篇论文发表在国内重要期刊和国际理
论研讨会论文集中，数篇曾获得优秀奖。出版有
《中国当代造型艺术家——刘娜纤维艺术作品集》
《匠心 纤手——当代纤维艺术形态的转变》。


Liu Jun
Shenzhen

Professor Vice-Director.Cultural & Creative
Products Institute of Shenzhen Polytechnic.
Bachelor of Arts (BFA), Suzhou University
Master of Arts (MFA) Wuhan University of
Technology
Visiting Scholar, University of Southampton,
UK
Expert Advisor,China National Silk Museum
Advisory Board
Board member of China Fashion
Association
Executive board member of International
Fashion Art Network
Vice Chairman, Committee of Shenzhen
Fashion Designers Association
Member of China Artists Association


Liu Peilin
Shenzhen

Gcraduated from Fashion design,
Shenzhen Polytechnic.
Honorable mention of the International
Undergraduate Art Award of the Year in
2020.
Double Bronze medal of Singapore Sands
Arts Festival in 2021.
In 2021, the works *START X START,
Interstellar Exploration* were seleeted by
the 4th Greater Bay Area College Art
and Design Works Exhibition, the 6th
Guangdong Province College Design
Works Biennale.
Bronze medal and Honorable mention of
the 3rd Taiwan Art Award in 2021.


Liu Na
Tianjin

Associate professor of Tianjin Academy
of fine arts, Member of Tianjin Artists
Association. Many works have been
selected into international and domestic
professional exhibitions. She has won a
bronze award and several excellent awards
in the International Fiber Art Biennale,
and have been invited to participate in
the World Tour of Chinese Contemporary
Fiber Art", and have participated in tour
exhibitions in Brazil, Poland, Colombia,
Ecuador, South Korea and other countries.
Many papers have been published in
the collection of papers in important
domestic journals and international
theoretical seminars, and several have won
excellent awards. He has published
*Chinese Contemporary Plastic artist-
Liu Na's collection of fiber art works* and
*Ingenuity· Slimmer-the Transformation of
Contemporary Fiber Art Form.*

刘韪
北京

清华大学美术学院设计学博士研究生。研究方向为服装艺术设计及理论研究。曾参与多项省、国家级课题，服装作品曾多次参与高校美术作品学年展等展览。

刘薇
北京

著名时尚艺术家
中国纺织非遗推广大使
高级工艺美术师、高级服装设计师
中国时装设计金顶奖设计师
紫禁城设计大奖获得者
玫瑰黛薇ROSEW品牌创始人、艺术总监

刘骧群
深圳

就职于深圳职业技术学院，艺术设计学院副教授，中国服装设计师协会会员，中国流行色协会会员。
2016年作品《甲》受邀参加韩国首尔服装艺术双年展。
2017年作品《烛》受邀参加"一带一路"沿线国家海内外艺术家服装艺术作品展，该作品被国家一级博物馆中国丝绸博物馆永久收藏。
2017年作品《烛》参加2017年度中国时装"匠·意时尚回顾展"。
2017年作品《纸甲》参加"文明的回响／致匠心"展览。
2018年4月在蒙德里安艺术基金会就荷兰现当代艺术、产品设计、服装设计等领域的科技发展问题进行了交流学习与研讨。
2019年作品《木兰》参加时装艺术国际特邀展并获得广东省大湾区高校教师作品展，二等奖。作品《纸甲》入围中国美术家协会主办的"复兴+再造"第十届中国现代手工艺学院展。
2021年，作品《藤甲–变色龙》参加"知者创物——第二届全国工艺美术作品展暨中国国家博物馆第二届工艺美术作品邀请展"。作品《致敬Iris van Herpen》，参加广东省大湾区高校教师作品展。

Liu Wei
Beijing

PhD candidate in theoretical study of design, Academy of Fine Arts, Tsinghua University. His research direction is fashion art design and theoretical study. He has participated in a number of provincial and national projects. His fashion works have participated in different exhibitions such as art and design exhibition of Chinese universities.

Rosew Liu
Beijing

Famous fashion artist
China Textile Intangible Cultural Heritage Promotion Ambassador
Master of Arts, Senior fashion designer
China Fashion Award China Fashion Design "TOP Award" winner
Winner of the Forbidden City Design Award
The Founder of ROSEW custom brand
Design director of Beijing Rosew Clothing Co., LTD

Liu Xiangqun
Shenzhen

Associate professor at Shenzhen Polytechnic, College of Art and Design, member of China Fashion Designers Association, and member of China Popular Color Association.
In 2016 The work Jia was invited to exhibit in the International Fashion Art Biennale in Seoul, Korea.
In 2017, the work Zhu was invited to exhibit in the exhibition of costume artworks by artists from home and abroad from countries along the "the Belt and Road", The work was permanently collected by the China National Silk Museum, a First Class National Museum in China.
In 2017, the work Zhu was exhibited in "To Craftsmanship—2017 Fashion Review Exhibition".
In 2017, the work Zhi Jia was exhibited in the exhibition "'The Silk Road and Celestial Clothes' Echo of Civilization" organized by the China Central Academy of Fine Arts.
In 2018, invited by Mondriaan Fund in Amsterdam, the Netherlands, to exchange study and discuss the development of technology in the fields of Dutch contemporary art, product design and fashion design.
In 2019 The work Mulan was exhibitedin the International Special Invitation Exhibition of Fashion Art.Thiswork won the Guangdong-Hongkong-Macao Greater Bay Area Educational Institution And The 4th Biennial Of The Academy Awards Of Guangdong Collage Design Works, second prize.
The work Zhi Jia was selected for the "Revival + Reengineering" 10th China Modern Handicraft University Exhibition sponsored by China Artists Association.
In 2021 the work Rattan Armor Chameleon was exhibited at the The Second National Arts and Crafts Exhibition and the Second Invitational Exhibition of Arts and Crafts Works in the National Museum of China.
The work Tribute to Iris van Herpen was exhibited in the exhibition of teachers' works in universities in Guangdong Province.

刘欣珏
中国澳门

毕业于英国中央圣马丁艺术设计学院和中央美术学院，分别学习面料设计和时装设计专业。SANCHIALAU的创始人及首席设计师；曾任时尚芭莎创新商务部时尚跨界艺术高级总策划；现任澳门奢侈品集团创意总监、澳门文化传播大使以及为沪澳青年交流促进会常务理事，中国服装设计师协会会员、中国广东设计协会会员设计师、粤港澳青年商会理事、珠海市海归青年交流促进会、澳门品牌商会会员，澳门菁英会、澳门侨界青年协会、澳门本土设计师协会会员。

刘寻
北京

媒体人、艺术家、策展人、北京织行合衣文化艺术传播有限责任公司创始人。曾就职中央电视台体育频道。2016年年初在雁巢收藏"织行合衣"作品发布展。2016～2019年先后分别在北京798的白盒子艺术馆和第零艺术馆举办刘寻"一根线"编织艺术展。

刘阳
西安

西安美术学院服装系副教授。2001年毕业于西安工程大学，服装艺术设计专业，获学士学位。2006年毕业于香港理工大学，服装及纺织专业，获硕士学位。专业方向：服装设计与表演、服装与服饰设计，曾荣获国家级、省级专业模特大赛、形象特使大赛冠、亚军和单项奖等奖项，担任专业赛事评委、拍摄平面及影视广告、参加大型服装表演展演。发表多篇学术论文，包括国家级核心期刊与CSSCI（中文社会科学引文索引），参与国家级、省级课题研究。积极参与社会实践，在专业中深入开展项目的实践与应用，服务社会。

Sanchia Lau
Macao, China

Designer Sanchia Lau completed her Bachelor degree in Fashion Design at China Central Academy of Fine Arts. And later she graduated from Central Saint Martins College of Art and Design, majoring in Textile Design. Now she is the founder and chief designer of the fashion & art brand SANCHIALAU, The senior planner of fashion art crossover project from the Creative Business Department of Harper's Bazaar, responsible for fashion photo shooting, stylish wardrobe design, fashion art crossover programs and organized activities including Harper's Bazaar Fashion Photo Shooting, Harper's Bazaar Tomorrowland Fashion Awards and other fashion art crossover events; And she is also the creative director of the magazine Real Estate Compass. Last but not least, she is one of the member designers of the Guangdong Design Association in China.

Liu Xun
Beijing

Media person, fasion designer, curator and founder of Beijing ZhiXing Heyi Culture and Art Communication Co Ltd. Took office in CCTV sports channel. In early 2016, the collection of weaving clothes was released in Yanchao. From 2016 to 2019, Liu Xun's One Thread Weaving Art Exhibition was held in Beijing 798 White Box Art Museum and Zero Art Museum respectively.

Liu Yang
Xi'an

Associate Professor of the Clothing Department at The Xi'an Academy of Fine Arts. She holds MA degrees in Textiles and Clothing Science at The Hongkong Polytechnic University, and BA degrees in Apparel & Art Design College at Xi'an Polytechnic University. Her professional directions include Fashion design and Performance, Fashion and Clothing design. She has won the awards of national and provincial professional model competition, Image Envoy Competition and etc. As a judge of professional competitions, shooting print and film advertising, and participating in many large-scale fashion shows. She also has published several journal papers including national core journals and CSSCI, and participate national and provincial research projects. She has actively participated in the social practice, carried out the practice and applications of projects intensively in her major to serve for the society.

刘一行
北京

引申品牌创意总监，一行一线国际时装教育机构创始人，毕业于中央民族大学服装设计系，毕业后曾去英国中央圣马艺术与设计学院进修，受邀京东平台在上海举办品牌动态时装周发布会。作品也曾多次在中国国际时装周举办个人时装发布会，作品曾被VOGUE、《芭莎》报道。

陆琰
上海

硕士，副教授，上海商学院服装与服饰设计专业系主任，主要从事服装与服饰设计的教学与研究工作，并致力于非物质文化遗产及传统服饰文化传承与活化设计。先后承担国家级精品资源共享课、国家精品课程等项目建设工作。主持并参与多项国家级、省级、校级科研项目。曾获大连杯国际青年设计师时装设计大赛银奖、益鑫泰中国服装设计最高奖优胜奖等荣誉。作品被中国丝绸博物馆收藏并展出。

罗杰
北京

四川美术学院设计学院服装与服饰设计专业教师、系主任助理，中央美术学院在读博士，时装艺术国际同盟副秘书长。获得教育部2019全国普通高等学校美术教育专业教师基本功展示个人全能一等奖;2017四川美术学院第二届教师教学技能竞赛"教学之星"称号，2018第四届重庆市高校青年教师教学竞赛二等奖；作品多次入选全国美展、重庆美展等专业展览并获奖，参与亚太经合组织（APEC）会议领导人服装设计、中国四川航空制服设计、四川美术学院新式学位服设计等项目，负责国家艺术基金资助项目及多部"十二五""十三五"规划教材编撰等科研和学术工作。

Liu Yihang
Beijing

Extended Brand Creative Director, graduated from the Department of Fashion Design at Minzu University of China, for YIHANGYIXIAN international fashion education institution. After graduation, he went to the Central Saint-Maarten School of Art and Design in the UK for further study and was invited to hold a brand dynamic fashion week conference in Shanghai on the JD.COM platform. His works have also held individual fashion shows at China International Fashion Week many times, and his works have been reported by *VOGUE* and *Bazaar*.

Lu Yan
Shanghai

Master, Associate professor, Head of the Department of Fashion and Costume Design, Shanghai Business School. She is mainly engaged in the teaching and research of costume and costume design, and is committed to the inheritance and activation design of intangible cultural heritage and traditional costume culture. It has undertaken the construction of national high-quality resource sharing courses and national high-quality courses. Host and participate in a number of national, provincial and university-level scientific research projects. She has won the silver award of Dalian Cup International Youth Fashion Design Contest, and the winning award of Yi Xintai China Fashion Design Highest Award. Her works were collected and exhibited in China Silk Museum.

Luo Jie
Beijing

Teacher and assistant to the head of the Department majoring in fashion and costume design in the design school of Sichuan Fine Arts Institute. A doctor in the Central Academy of Fine Arts. Deputy Secretary General of the International Fashion Art Network.Won the first prize of the Ministry of education's 2019 individual all-round exhibition of basic skills of teachers majoring in art education in national colleges and universities. The title of "teaching star" in the second teachers' teaching skills competition of Sichuan Fine Arts Institute in 2017. The second prize in the fourth young teachers' teaching competition of colleges and universities in Chongqing in 2018. His works have been selected into national art exhibition, Chongqing Art Exhibition and other professional exhibitions for many times.He has participated in the fashion design of leaders of APEC conference, the uniform design of Sichuan Airlines, the new degree dress design of Sichuan Academy of fine arts and other projects, and is responsible for the scientific research and academic work such as the projects funded by the national art foundation and the compilation of several teaching materials.

罗娟
南充

毕业于江南大学设计艺术学硕士,清华大学访问
学者。西华师范大学美术学院服装与服饰设计专
业教师。
研究方向为红色文化服装设计。紧抓红色文化脉
搏,以革命、抗战为主题进行设计创作。

罗莹
深圳

教授,深圳大学艺术学部美术与设计学院副院
长,硕士生导师。
兼任中国服装设计师协会学术委员会执行委员,
中国纺织服装教育学会产学研委员会秘书长,英
国伯明翰城市大学艺术设计学院访问学者;
时尚植物染品牌"草木蓝兮"创始人,"无废植
染"联合创始人。

罗峥
深圳

中国服装设计最高奖第10届金顶奖获得者,中
国著名女装设计师,APEC会议领导人服装主
创。中国著名女装设计师品牌"OMNIALUO欧
柏兰奴""LuoZheng华服"创立人;深圳东方
逸尚服饰有限公司董事长兼艺术总监;深圳市时
尚文化创意协会会长;深圳市政协常委;深圳市
第八届工商联副主席。

Luo Juan
Nanchong

Graduated from Jiangnan University,
Master of Art design, Tsinghua University
Study as a visiting scholar. Teacher of
Fashion Design, West China Normal
University. Research direction is Red
cultural clothing design.

Luo Ying
Shenzhen

Professor vice dean, and master director
of College of Fine Arts and Design, Faculty
of Arts, Shenzhen University.
Executive member of academic
Committee of China Fashion Designers
Association.Secretary General of industry-
University-Research Committee of China
Textile and Garment Education Society.
Visiting Scholar, College of Art and Design,
Birmingham City University, UK.
Fashion plant dyeing bzand founder of
"Indigo In", co-founder of "Zero- waste
plant dyeing".

Cindy Luo
Shenzhen

A well-known designer in China, she was
the winner of the 10th. Jing Ding Award,
which is the Orscar in Costume Designing
in China. She is main dexigner leaders and
first ladies in APEC 2014. Now she is the
oufnder of two brands "Omnialuo" and
"LUOZHENG" Art Director of Shenzhen
Oriental Fashion Co,.Ltd; She is also the
Chairman of Shenzhen Fashion Culture
Creative Association, member of Shenzhen
CPPCC Standing Committee, vice
chairman of the 8th session of Shenzhen
Federation of industry and Commerce
(General Chamber of Commerce).

吕越
北京

教授、设计师、艺术家、策展人。中央美术学院时装设计创建人，学科带头人。中国美术家协会会员，中国美术家协会服装艺委会成员。时装艺术国际同盟主席。中国纺织工程学会时装艺术委员会主任，中国服装设计师协会学术委员会专家委员，中国流行色协会理事。时装艺术国际展策展人。
她在时装设计和艺术创作上均有造诣。曾获得过多个奖项，多次出任国际时装比赛的评委，多次受邀与相关机构合作。她的艺术作品和时装设计作品在多个展览以及活动中展出，展出地点包括：西班牙、丹麦、意大利、泰国、韩国、土耳其、法国、日本、美国、挪威、中国等。作品被多家机构和个人收藏。

穆罕默德·法瓦德·努里
巴基斯坦

现为巴基斯坦GIFT大学美术、设计与建筑学院助理教授。最近被任命为艺术与设计学院国际教师协调人，该学院是斯里兰卡第一所与意大利罗马Cordella学院合作的国际设计学院。他曾获得SDC巴基斯坦设计顾问服务铜奖；2020年荣获"Odartey"风格和时尚奖评选的时尚偶像成就奖；匈牙利国家时装联盟国际执行理事会成员及GSFW全球可持续时装周创意设计总监。

牟琳
北京

硕士毕业于中央美术学院设计学院艺术设计专业；
2006，《秋天的童话》，"以纸为料"中央美术学院与韩国弘意大学合作艺术展；
2007，《青衣之恋》入选2007国际时装设计师大奖赛总决赛；
2008，《绽放》，北京国际时装艺术展；
2012，《展开的线团儿一》，游园-2012北京国际时装艺术展，作品在嘉德秋拍拍卖；
2013，《展开的线团儿二》，日日夜夜-2013北京国际艺术展。

Lyu Yue (Aluna)
Beijing

Designer, artist and curator. Professor of Central Academy of Fine Arts( CAFA). Founder to Fashion Department, CAFA. Member of Chinese Artists Association, Member of Costume Art Committee of Chinese Artists Association. President, International Federation of Fashion Arts. Director of fashion Art Committee of China Textile Engineering Society. Expert member of academic Committee of China Fashion Designers Association Director of China Popular Color Association. Curator of international Exhibition of Fashion Art. She is accomplished in fashion design and artistic creation. She has won many awards, served as a judge of international fashion competitions for many times, and has been invited to cooperate with relevant organizations for many times. Her artwork and fashion designs have been exhibited in several exhibitions and events, exhibition venues include: Spain, Denmark, Italy , Thailand, South Korea, Turkey, France , Japan , the United States, Norway, China. Her works are collected by many institutions and individuals.

Muhammad Fawad Noori
Pakistani

Currently working as Assistant Professor at GIFT University in School of Fine Arts, Design and Architecture. Recently appointed as Coordinator of the International Faculty at Academy of Art & Design School (AND) is Sri Lanka's first international design Academy with partnership of Instituto Cordella Roma, Italy. BRONZE MEDAL AWARDED BY SDC UK Bradford on enthusiastic services for SDC Pakistan as design consultant.
Winner of style icon achievement award in Odartey style and fashion awards 2020. International executive board member of National Fashion League Hungarian association and Creative design Head of GSFW global sustainable fashion week since 2016.

Mu Lin
Beijing

Master of Artistic Design, School of Design, Central Academy of Fine Arts.
International Exhibition
2006, *An autumn's tale*, Paper or Tissue Cooperative Fashion Art Exhibition between the China Central Academy of Fine Arts and Korea Hong-IK University
2007, *Tsing yi of love*, the finals of International Fashion Designer Competition
2008, *Blooming*, From Where Beijing Fashion Art International Exhibition.
2012, *Expanded Coil One*, the Game Home 2012 Beijing Fashion Art International Exhibition
Auctioned by China Guardian International Auction co.,LTD
2013, *Expanded coil Two*, the Day &Night 2013 Beijing Fashion Art International Exhibition

农琼丹
南宁

广西艺术学院设计学院服装与服饰专业专任教师，中级工艺美术师；
2002—2006年就读广西艺术学院 美术师范系美术教育专业；
2011—2015年就读广西艺术学院学院设计学院民族传统服饰与服装艺术专业；
2016年，《黑白同色》获第十八届全国设计"大师奖"服装设计类金奖；
2015年，《几何原色》获"迪尚"第十届中国时装设计大赛成果奖评选入围奖；
2014年，《以梦为马》获得第八届"浩沙杯"国际泳装设计大赛铜奖；
2016年，《梦境》获穿越这一刻：中英文化交流年——全球"中国风"男装设计 大奖赛最具商业潜力大奖；
2006年《游乐空间》获"浩沙杯"第三届国际泳装设计大赛优秀奖。

Nong Qiongdan
Nanning

Full time teacher of clothing and apparel major at Design College of Guangxi Academy of Arts intermediate craft artist.
2002–2006 art education major of art normal Department of Guangxi Academy of Arts
2011–2015 major in ethnic traditional clothing and fashion art, School of design, Guangxi Academy of Arts.
In September 2016, *Black and white* in the same color won the gold award of fashion design of the 18th National Design "Master Award".
In October 2015, *Geometric primary color* won the Finalist Award of the 10th China Fashion Design Competition of "Dishang".
In March 2014, *Taking dreams as horses* won the bronze medal in the 8th "Haosha Cup" international swimsuit design competition.
In 2016, *Dream* won the most commercial potential award of the global "Chinese style" men's wear design Grand Prix, the year of Sino British cultural exchange.
In 2006, *Amusement space* won the excellence award of the third international swimsuit design competition of "Haosha Cup".

潘璠
西安

西安美术学院服装系教授、硕导。
研究方向为生态纺织服装绿色设计及应用，担任教育部人文社会科学研究青年基金项目、陕西省教育厅科研项目、校级艺术扶贫项目负责人，申请多项专利。出版专著《手绘服装款式设计与表现》、"十三五"普通高等教育本科部委级规划教材《生态纺织服装绿色设计》《电脑艺术时装画》，参编"十二五"等教材。在cssci及核心期刊发表多篇论文。服装服饰艺术设计作品多次入选全国及省级美术作品展览。

Pan Fan
Xi'an

Professor and master supervisor of the department of clothing in Xi'an Academy of Fine Arts.
Her research direction is green design and application of ecological textile and clothing. She served as the director of the youth fund for humanities and social sciences research of the Ministry of Education, the scientific research project of Shanxi Provincial Department of education and the art poverty alleviation project at the university level, and she applied for many patents. She published a monograph Design and Demonstration of *Hand-drawn Clothing Styles*, *Green Design of Ecological Textile* and *Clothing, Computer Art Fashion Drawing*, and participated in the compilation of textbooks such as the ones for the 12th Five Year Plan. She has published many papers in CSSCI and key journals.Her clothing design works have been selected into national and provincial art exhibitions for many times.

潘静怡
中国香港

多媒体艺术家。毕业于澳洲皇家墨尔本理工大学(艺术系硕士)，英国赫尔大学(工商管理硕士)。曾受委托的艺术创作项目包括:香港交易所—为"深港通"创作代表香港一方的铜牛纪念雕塑；为连卡佛Joyce集团总部设计及装置两台不锈钢雕塑作品；房屋署苏屋村重建-以苏屋旧物设计及装置两个单位；香港恒生大学——以扩增实境"AR"新媒体创作公共艺术作品，体现我们独特的中国文化传统及文化遗产；以及希慎兴业有限公司赞助2019法国五月的联办节目"华丽变奏"个人展览。现为香港理工大学服装馆委员会的成员，亦于2020年为理大纺织及制衣学系毕业展的评审，作品《天圆地方》参展杭州中国丝绸博物馆举办的2019国际旗袍展览。

Helen Pun
Hongkong, China

Banker-turned artist, she holds a Master of Fine Art degree from the RMIT University, Melbourne (with distinction); and an MBA degree from the University of Hull, UK (with distinction). She has been commissioned a number of art projects, for example by the Hongkong Exchanges (a bull bronze sculpture for "Shenzhen Connect"), Housing Authority (So Uk Redevelopment installation), Lane Crawford Joyce Group (sculptural installation), Hang Seng University of Hongkong (augmented reality public art project) and Wheelock Development (sculptures for a new development site in Kai Tak). As a fashion artist, she was sponsored by Hysan Development Co. Ltd. for a Le French May 2019 project (Vivre la Cheongsam) at Lee Garden One; also by Hongkong Polytechnic University for a solo exhibition in 2019. She was a judge for PolyU'sgraduation show in 2020 and now a member of the Gallery Academic Committee of The Fashion Gallery of PolyU. Helen's fashion artwork ("Harmony"), representing Hongkong (PRC), was featured at the 2019 Global Qipao exhibition held at the China National Silk Museum in Hangzhou

彭青歆
中国香港

香港理工大学博士后。她的研究领域为时尚传播与推广，主要将创新性纺织品应用于服装设计中，并以线上和线下等活动形式进行服饰及品牌推广。

秦瑞雪
重庆

硕士，毕业于四川美术学院服装与服饰设计专业，现工作于四川美术学院，主要研究方向为"可穿着艺术"。获全国中小学校服设计大赛铜奖，设计作品先后入围重庆美展、天然染料双年展、中国真维斯杯休闲装设计大赛、中国马海毛时装创意设计大赛、紫金奖中国大学生设计展等展览和比赛。主持、主研重庆市艺术科学研究规划等省级项目2项、四川美术学院研究生科研创新项目等校级项目3项，以及四川美术学院校级重大项目1项。论文发表于CSSCI来源期刊《装饰》1篇，省级期刊1篇。

邵芳
北京

北京城市学院艺术设计学部服装与服饰教研室设计主任，硕士研究生导师；
1995—1999年就读于中央工艺美术学院染织服装系，获文学学士；
2007—2008年就读于意大利马兰欧尼时装学院时装设计专业，获硕士研究生。

Peng Qingxin
Hongkong, China

Postdoctoral fellow in The Hongkong Polytechnic University. Her research area is fashion communication and promotion including the application of creative textiles in fashion designs and promoting the fashion and brand in on/offline environments.

Qin Ruixue
Chongqing

Master degree, graduated from Sichuan Academy of Fine Arts, majoring in clothing and apparel design, now works in Sichuan Academy of Fine Arts, and his main research direction is "wearable art". Won the Bronze Award of the National Primary and Secondary School Uniform Design Competition, and his design works have been shortlisted for Chongqing Art Exhibition, Natural Dyestuff Biennale, China Jeanswest Cup Casual Wear Design Competition, China Mohair Fashion Creative Design Competition, and Purple Gold Award China College Student Design Exhibition and other exhibitions and competitions. Presided over and researched 2 provincial-level projects such as Chongqing Art Science Research Planning, 3 school-level projects such as the Sichuan Academy of Fine Arts Graduate Research and Innovation Project, and 1 major school-level project of Sichuan Academy of Fine Arts. The paper was published in 1 article in the source journal "Decoration" of CSSCI and 1 article in the provincial journal.

Shao Fang
Beijing

Head of the Department of Fashion Design / Postgraduate tutor, Beijing city university.
1995–1999, BA, Central Academy of Arts and Crafts Department of Fashion Design
2007–2008, WA, Institute Malangoni in Italian, Fashion design.
Master's Degree

石历丽
西安

西安美术学院服装系教授、硕士研究生导师，中央民族大学民俗学博士，中国服装设计师协会学术委员会理事，陕西省服装行业协会专家委员会委员。参展经历：时装艺术展、全国纺织品设计大赛、纤维艺术展等。

石梅
北京

毕业于解放军艺术学院音乐系，总政歌舞团任演奏员、北京教育学院音乐系教师。作品《韵》参加2015年第七届国际时装艺术展，作品《佳人有约》参加2016年第八届国际时装艺术展，作品《和远古有个约会》参加2018年第九届国际时装艺术展，作品《心灵赋格》参加2018年"从洛桑到北京"第十届国际纤维艺术双年展，作品《千年对话》参加千年对话——2019时装艺术国际特邀展，作品《岁月留痕》参加新世界，新视界——2020时装艺术线上国际展，作品《驶向远方》参加"共生共存"—第十一届"从洛桑到北京"国际纤维艺术双年展作品《侘》《寂》参加2021年"唯物思维"中国—乌拉圭当代材料艺术交流展（线上展览）2021年，作品《中国"她"艺术》大型艺术刊物。

覃莉方
南宁

2006年毕业于北京服装学院服装艺术设计专业，获学士学位。2007年毕业于香港理工大学服装与纺织品设计专业，获得硕士学位。
2011年，作品《黑与白》获第二届广西女性书画摄影手工艺作品展（美术类）三等奖；
2013年，作品《瞬（1.2.3）》获第四届广西女性书画作品展（美术类）二等奖；
2017年，年作品《瓷锦印象》获2017广西艺术作品展优秀奖；
2018年，作品《路图》入选首届全国工艺美术作品展；作品《银装》获首届'纺织之光'中国纺织艺术展优秀奖；作品《丝影》获十二届"中国—东盟青年艺术创作大赛"优秀奖；
2019年，作品《草木江山》获2019广西艺术作品展优秀奖；
2020年，作品《防护内外》入选第三届中国时装画大展。

Shi Lili
Xi'an

professor of clothing department of Xi'an Academy of Fine Arts, master tutor, Minzu University of China folklore Ph.D., the academic committee China clothing designer association director, member of Shaanxi Province garment industry association expert committee. Fashion art exhibition, the national textile design contest, fiber art exhibition.

Shi Mei
Beijing

Graduated from the People's Liberation Army Academy of Art. As a performer in the General Political Department Song and Dance Troupe and a music teacher in Beijing Institute of Education.
Work Rhythm was displayed in the 7th edition International Fashion Art Exhibition 2015.
Work An Appointment of a Beauty was displayed in the 8th edition International Fashion Art Exhibition 2016.
Work Have an appointment with ancient times was displayed in the 9th edition International Fashion Art Exhibition 2018.
Work A Date with Distant Past was displayed in the 11th From Lausanne to Beijing International fiberart Biennale Exhibition.
Work Millennium dialogue was displayed in the Millennium dialogue edition International Fashion Art Exhibition 2019
The work Traces of years was displayed in the new world, new vision-2020 fashion art online international exhibition
The work Sailing far away was displayed in the "symbiosis and coexistence"- the 11th International Fiber Art Biennale "from Lausanne to Beijing".
His works Yu and Ji was displayed in the "materialistic thinking" China Uruguay contemporary material art exchange exhibition (online exhibition) in 2021.
In 2021, her work Chinese art was published as a large art publication.

Qin Lifang
Nanning

In 2006, she earned the bachelor's degree in fashion design, Beijing Institute Of Fashion Technology.
In 2007, she earned the master's degree in fashion and textile design, The Hongkong Polytechnic University.
In 2011, work Black and white won three prize in the second session Guangxi women's calligraphy and painting photography and Handicraft Exhibition.
In 2013, work Instant (1.2.3) won two prize in the fourth Guangxi women's painting and calligraphy exhibition.
In 2017, work Impression of porcelain brocade won the excellence award of 2017 Guangxi art exhibition.
In 2018, work Road map was selected into the first national arts and crafts exhibition. Work Silver dress won the excellence award of the first "light of textile" China Textile Art Exhibition. Work Silk shadow won the excellence award of the 12th China ASEAN Youth art creation competition.
In 2019, work Vegetation, rivers and mountains won the excellence award of 2019 Guangxi art exhibition.
In 2020, the work Protection inside and outside was selected into the third Chinese fashion painting exhibition.

汪子丁
北京

2017—2021 本科就读于北京服装学院；
2018年作为志愿者参与北服承办的IYDC国际设计邀请赛，协助英国设计师获得金奖；
2020年参与BIFT-MMU国际工作营项目，设计被选入优秀作品，获得Victoria Beckham品牌面料赞助，参与时尚电影拍摄；
2021年入围美国AOF青年设计师大赛；
2021年以总分第一的成绩考入清华大学美术学院染织服装设计系服装设计专业，攻读研究生学位。

王丹丹
西安

西安美术学院服装系教师。研究方向为服装与服饰艺术设计。中国服装设计师协会学术委员会会员；陕西省青年联合会第十二届委员会委员；西安市青年联合会第八届委员会委员。发表专业论文10余篇，个人作品《梦青铜》《晨暮中的穿行者》《秦绣 红剪》《墨荷》《夏禅》《记忆碎片》《玄观》《楚 龙凤虎》等先后入选参展并获得重要奖项。

王雷
北京

2010年毕业于中央美术学院造型学院实验艺术系，获艺术硕士学位。系中国美术家协会会员、中国雕塑学会会员、中国纺织工程学会时装艺术专业委员会副主任、河南省文联第八届委员会委员。现为浙江师范大学"双龙学者"特聘教授、硕士生导师、美术学系主任。曾在中国美术馆、河南省美术馆、81美术馆等单位举办个展8次，自2007—2020年参加国内外学术展览180余次，其中国家级展览30余次（5件作品入选第12、13届全国美展）。曾获第二届"中国美术奖·创作奖·铜奖"等奖项。作品《文化中国》被国家艺术基金创作扶持，《寻找红五星》《新长征路上的凯歌》被中国文学艺术发展专项基金扶持。作品被中国美术馆、南京博物院、澳大利亚白兔美术馆等20余家单位及中（大陆、港、台）、美、英、德、意、西、澳、韩、瑞士、新加坡等十余个国家私人收藏。

Ella Wang
Beijing

From 2017 to 2021, studied in Beijing Institute of Fashion Technology as an undergraduate and was admitted to the experimental class of Fashion and Costume Design.
In 2018, she participated in the IYDC International Design Invitational held by BSC as a volunteer and assisted British designers to win the gold medal.
In 2020, she participated in bift-MMU International Work Camp project, and my design was selected as excellent work. She was sponsored by Victoria Beckham and participated in fashion film shooting.
In 2021, she was selected for AOF Young Designers Competition in the United States.
In 2021, she was admitted to the Fashion Design major of Dyeing and Weaving Fashion Design Department of Academy of Fine Arts of Tsinghua University with the highest score, and studied for a postgraduate degree.

Wang Dandan
Xi'an

Teacher of Fashion Department of Xi'an Academy of Fine Arts. Research direction is clothing and apparel art design. Member of the Academic Committee of China Fashion Designers Association; member of the 12th Committee of Shaanxi Youth Federation; member of the 8th Committee of Xi'an Youth Federation. Published more than 10 professional papers, personal works *Dream Bronze, Walker in the Morning and Evening, Qin Embroidery· Red Scissors, Ink Lotus, Xia Chan, Memory Fragments, Xuan Guan, Chu Dragon, Phoenix and Tiger*. She has been selected to participate in the exhibition and won important awards.

Wang Lei
Beijing

Professor and Chair of the Fine Arts Department at Zhejiang Normal University in China. He has held eight solo exhibitions including National Art Museum of China, Henan Art Museum, Eight One Art Museum and others. From 2007 to 2020, he has participated in more than one hundred seventy academic group exhibitions nationally and internationally. His works were selected for the 12th and 13th National Exhibition of Fine Art in China and won Bronze Award of the 2nd "Chinese Art Award, Creation Award". His projects *Culture China* was Supported by China National Arts Fund as well as *Looking for Red Five Star* and *Triumphant Song on the New Long March* projects were supported by China Literature and Art Foundation. Wang Lei's artworks are collected by many Chinese museums and galleries including National Art Museum of China, Nanjing Museum,White Rabbit Collection and so on more than 20 units and China (mainland, Hongkong, Taiwan), the United States, Britain, Germany, Italy, Spain, Australia, Korea, Switzerland, Singapore and so on more than 10 countries private collection.

王妮
武汉

武汉纺织大学服装学院副教授，硕士研究生导师。教育部高等学校青年骨干教师国内访问学者，中国纺织服装教育学会产学研委员会委员，中国蓝染产业智库专家，深圳市服装设计专业高级职称评审委员会专家。研究方向：荆楚传统印染传承与创新，荆楚纺织非遗文化资源与文化产业。主持省级社科重大、一般、文旅部重点实验室项目共7项，"纺织之光"中纺联教学改革研究项目2项。参与国家社科重点项目1项，国家艺术基金1项，文旅部非遗研培计划6项，指导国家级和省部级大创项目获奖多项。十五年长期专注于《创意手工印染》课程教学。出版教材专著3部，发表学术论文（含核心作品）30余篇，设计作品获奖20余次，国家级与省部级行业内展览10余次。

王培娜
青岛

硕士生导师，青岛大学副教授，广东省服装设计师协会副会长，中国服装设计师协会时装艺术委员会委员，青岛女装发展专家咨询委员会委员，国际FD职业设计师学会终身会士，高级服装设计师，中国十佳时装设计师。

王文
广州

香港理工大学时尚与纺织品设计博士，广州美术学院副教授。曾在英国皇家艺术学院博士交流学习。研究成果发表于多个国际会议和学术期刊。主要研究方向：时尚与纺织品创新设计，可持续未来设计和设计思维等。曾荣获中国设计师新人奖，多次参与香港时装周和伦敦时装周，个人作品入选世界当代纺织艺术双年展。

Wang Ni
Wuhan

A brief Introduction of Wang Ni in the 2021 edition (Simplified)
Wang Ni, associate professor in School of Fashion of Wuhan Textile University, postgraduate tutor , Visiting Scholar for Young Key Teacher of Higher Education of The Ministry of Education, a committee of the CTAES(China Textile and Apparel Education Society) ,an expert think tank of China Blue Dyeing Industry, an expert of Senior Professional Title Appraisal of Shenzhen Fashion Design.
Research Area: Inheritance and innovation of Jing-chu Traditional Printing and Dyeing, Jing-chu Textile Intangible Cultural Resources and Cultural Industry.
Scientific Research Project:
7 provincial Key Laboratory projects of major general and cultural Tourism Department of social sciences;
2 teaching reform research projects of textile association of "Textile Vision";
1 state key project in social sciences ;
1 national Arts fund;
6 intangible cultural research and cultivate programs of Ministry of Culture and Tourism ;
An advisor in national and provincial College Students Entrepreneurship Programs ;
Teaching course: Creative Hand Dyeing
Public:
3 published textbooks and monographs;
More than 30 published academic papers (including core journal);
More than 20 awards design works;
Activity: More than 10 nationals and provincial professional exhibitions.

Wang Peina
Qingdao

Master tutor, associate professor of Qingdao University, vice president of Guangdong Fashion Designers Association, member of Fashion Art Committee of China Fashion Designers Association, member of Qingdao Women's Wear Development Expert Advisory Committee, life member of FD Professional Designers Association, Senior Fashion Designer, Awarded  China's top 10 fashion Designers.

Wang Wen
Guangzhou

PhD in Fashion and Textile Design, The Hongkong Polytechnic University. Associate professor at Guangzhou Academy of Fine Arts. She has studied at the Royal College of Art, London for doctoral exchange. Her research has been published in several international conferences and academic journals. Research interests include innovative design in fashion and textiles, sustainable future design, and design thinking, etc. She has won the "China New Designer Award", and participated in Hongkong Fashion Week and London Fashion Week several times. Her artworks were selected for the World Contemporary Textile Art Biennale.

郑萌泽
北京

中央美术学院学生，2019 年参与北京国际时装周，2020年参展上海中华艺术宫全国高等美术教育展，多次与国内外杂志媒体博主等合作。

Zheng Mengze
Beijing

Student of The Central Academy of Fine Arts (CAFA). Participated in Beijing International Fashion Week in 2019. 2020 Shanghai China Art Museum National Higher Art Education Exhibition.Many times with domestic and foreign magazine media bloggers and other cooperation

王钰涛
北京

中国服装设计师协会兼职副主席、常务理事，青岛大学客座教授，北京服装学院校外研究生导师。
2011年荣获梅赛德斯－奔驰中国国际时装周先锋设计师，荣获第15届中国服装设计师最高奖"金顶奖"，荣获中国纺织服装行业十大设计师及年度创新人物称号
2012年获邀参加梅赛德斯－奔驰德国柏林时装周2012 A/W专场发布。
2014年荣获年度哥本哈根皮草设计大师奖，荣获亚太经合组织（APEC）会议人领导人服装样衣制作工作并作出突出贡献奖。
2015年获邀参加香港华丽秀发布2015 A/W女装系列发布，受邀参加澳大利亚悉尼时装周2016 S/S专场发布。
2017年荣获第21届中国服装设计师最高奖"金顶奖"，自创品牌B+荣获年度时尚品牌奖，第二届耀动华人颁奖盛典 年度杰出华人设计师。
2018年荣获中国国际时装周"最佳女装设计师"时尚大奖。
2020年自创品牌Beautyberry荣获年度时尚品牌奖。受邀参与录制CCTV-3综艺频道《回声嘹亮》，受邀参与录制CCTV-3综艺频道《衣尚中国》，荣获北京2022冬奥会和冬残奥会制服装备视觉外观设计征集评审活动优秀奖。

Wang Yutao
Beijing

Part-time Vice Chairman and Executive Director of China Fashion Association. Visiting Professor of Qingdao University, External. Postgraduate Supervisor of Beijing Institute of Fashion Technology.
In 2011, awarded Avant-garde designer Mercedes-Benz China Fashion Week. Awarded the highest award Jinding Prize in the 15th China Fashion Designer, Awarded top 10 designers and annual innovation figure of China textile and garment industry.
In 2012, invited to participate in Mercedes-Benz Berlin Fashion Week  and launched 2012 A/W special show.
In 2014, awarded annual Kopenhagen Fur Design Master Award,Conducted sample production for APEC leaders clothing and awarded Outstanding Contribution Award.
In 2015, invited and launched 2015 A/W Womenswear Collection in Hongkong Fashion Extravaganza, Invited and launched 2016 S/S special show in Australia Sydney Fashion Week.
In 2017, awarded the highest award Jinding Prize in the 21st China Fashion Designer, self-brand B+ awarded Fashion Brand of the Year in 2017 Awarded Outstanding Chinese Designer of the Year at the 2nd Shining Chinese Awards Ceremony.
In 2018, awarded Fashion Award "Best Womenswear Design" of China Fashion Week.
In 2020, Self-brand Beautyberry awarded Fashion Brand of the Year,Invited to participate in the recording of CCTV-3 Variety Channel *Loud and Clear Echo*, Invited to participate in the recording of CCTV-3 Variety Channel *Clothing and Fashion in China*. Awarded Excellent Award of uniform visual design collection and evaluation activities for Beijing 2022 Olympic Winter Games and Paralympic Winter Games.

王悦
北京

清华大学美术学院染织服装系长聘副教授、博士生导师；清华大学美术学院哥本哈根皮草实验室主任；英国伦敦时装学院、中央圣马丁艺术设计学院访问学者；中国流行色协会色彩教育专业委员会委员、时装艺术国际同盟学术委员会理事、IFTF国际毛皮协会青年委员会理事、北京高等院校人物造型设计教学联盟专家委员。致力于服装艺术设计的可持续发展与传统服饰文化研究，研究方向包括服装艺术设计与传统服饰文化研究、材料与服装设计及可持续发展研究。

Wang Yue
Beijing

Associate professor with tenure of Academy of Arts & Design, Tsinghua University; Director of Tsinghua-Kobenhagen Fur Studio; Visiting Scholar of University of the Arts London / London College of Fashion; Committee Member of Professional Committee of Color Education, China Fashion Color Association; Director of Academic Committee of Fashion Art International Alliance; Expert Committee Member of Teaching Alliance of Character Image Design of Beijing Institute of Higher Education. Research Directions related on fashion design and traditional fashion culture & handicrafts; Sustainable development on materials and fashion design.

王志惠
长春

吉林艺术学院服装设计系系主任，副教授，硕士研究生导师，2014年获得第22届"汉帛奖"国际青年时装设计师大奖赛优秀奖。

吴帆
深圳

纺织与服装学者、当代纤维艺术家，深圳职业技术学院艺术与设计学院服装系教授，中国美术家协会会员，中国工艺美术协会纤维艺术专业委员会副主任，深圳美术家协会美术教育专业委员会副主任。
作品曾参加造型艺术展、全国美展、中国当代纤维艺术展、"从洛桑到北京"国际 纤维艺术双年展、国际绞缬艺术展等重要展事并获奖，作品曾受邀赴美国、德国、丹麦、立陶宛、波兰、墨西哥、意大利、英国、澳大利亚、格鲁吉亚、日本等国家参加重要群展。

吴晶
成都

硕士生导师，四川大学轻工科学与工程学院副教授；中国服装设计师协会学术委员会委员；香港理工大学纺织与制衣系硕士；英国南安普顿大学温彻斯特艺术学院访问学者。服装设计作品曾在中国、美国、韩国、中国香港、中国台湾等国家和地区展出。

Wang Zhihui
Changchun

Director of fashion Design Department, Jilin university of the Arts. Associate professor,Postgraduate tutor.Excellent Award of Hempel Award the 22nd China International Young Fashion Designers Contest.

Wu Fan
Shenzhen

Chinese contemporary fiber artist, fashion designer, scholar and curator. Works bridge fashion, textiles, installation and sculpture. From 1987 to 1992, she studied textiles, dyeing and weaving design and fashion design at School of Textiles, Dong Hua University and Academy of Arts & Design , Tsinghua University. From 1999 to 2006, She obtained an MA in School of Art and Design, Wuhan University of Technology and continued to being a visiting scholar at Hongkong Design Institute(HKDI) and School of Art and Design Auckland University of Technology(AUT). She is a professor of School of Art and Design, Shenzhen Polytechnic. Her artworks were exhibited in 2nd 11th From Lausanne to Beijing International Fiber Art Biennal Exhibition in China; Contemporary Chinese Fiber Art Exhibition in San Jose Museum of Quilts & Textiles, etc.; the 2nd Modeling Art Exhibition of New Artists in National Art Museum of China, the 12th National Exhibition of Fine Arts; Fiber Art by Wu Fan in Shenzhen Guan Shan Yue Art Museum; 8th International Contemporary Textile Art Biennale WTA in The Centro Cultural Galileo in Madrid, etc. She was invited to participate in important group exhibitions in USA, Germany, Denmark, Lithuania, Poland, Mexico, Italy, UK, Australia, Georgia, Japan and other overseas countries.

Wu Jing
Chengdu

Associate professor of Sichuan University supervisor of postgraduates master of textile and garment in Hongkong Polytechnic University. Visiting scholar in University of Southampton Winchester School of Art. Member of Academic Committee of China Fashion Designers Association. Her fashion design works have been exhibited in China, the United States, South Korea and other countries and regions.

吴巧玲
金华

浙江师范大学美术专业硕士研究生

伍秋裕
北海

任职于北海艺术设计学院服装专业专任教师，毕业于广西艺术学院艺术设计硕士研究生，研究方向为传统民族服饰与服装艺术。作品2019年入选第十三届全国美术作品展；2021年获"迪尚杯"第14届新生代设计大赛金奖；2020年获"敦礼为衣"——第四届柯桥·中国国际礼服设计大赛金奖；2020年获"大浪杯"中国女装设计大赛银奖及网络人气奖；2020年入选《众志成城设计抗疫——同心抗疫设计作品选》；2020年获第二十三届"真皮标志杯"效果图优秀奖等17个作品比赛奖项；2019年论文《白裤瑶粘膏画染研究》发表于《西部皮革》ISSN1671–1602。2020年论文《侗琵琶在文创产品设计中的运用与研究》发表于《民博论丛》。

萧颖娴
杭州

艺术理论与实践博士，自2021年开始作为独立艺术家和设计师工作生活于杭州和加拿大，现为加拿大康考迪亚大学MILIEUX艺术研究所纺织与材料学会会员，研究方向为纺织服装材料的可持续设计：一种创造性的纺织品设计方法，用于设计材料组合的可拆卸连接。2005至2021年任中国美术学院染服系服装设计专业教师，主持完成省部级课题，参与完成多个国家级课题。艺术时装作品多次入选国内外的艺术展览，其中包括上海当代美术馆和广州当代美术馆的艺术展，北京国际艺术时装展，以及韩国、德国、荷兰和意大利等国际艺术时装与设计展。艺术时装作品多次入围全国美展并获获奖提名，且作品多次在浙江省美展中获奖。

Wu Qiaoling
Jinhua

Postgraduate Student, in Fine Arts,
Zhejiang Normal University.

Wu Qiuyu
Beihai

She works as a full-time teacher in Costume major of Beihai University of Art and Design. She graduated from Guangxi University of Art and Design with a master's degree in art design. Her research direction is traditional national costume and costume art. In 2021, won the gold medal of the 14th New Generation Design Competition of "Dishang Cup"; In 2020, it won the "Dun Li for Clothes" — the 4th Keqiao. Gold Medal of China International Dress Design Competition; In 2020, silver award and Network Popularity Award of "Dalang Cup" China Women's Design Competition; In 2020, it was selected as "Zhongzhi Chengcheng Design anti-Epidemic — Concentric Anti-Epidemic Design Works"; In 2020, won the 23rd "Genuine Leather Logo Cup" effect drawing excellence Award and other 17 work competition awards; 2019 paper *Research on Painting and Dyeing of Baitrousers Yao Adhesive Paste* published in Issn1671-1602. In 2020, the paper *"Application and Research of Dong Dong Pipa in Cultural and Creative Product Design"* was published in Minbo Collection.

Xiao Yingxian
Hangzhou

Ph. D in art theory and practice.
From 2021, she work and live in hangzhou and monteal as an independent artist and designer. She is also the member of textile & materiality research cluster of milieux institute of arts, culture and technology at concordia university, her research is focus on sustainable design of textile: a creative textile design methodology for designing detachable connections for material combinations.
From 2005 to 2021, she served as professor in fashion department of china academy of art, presided over the completed provincial and ministerial-level research projects, and participated in many national-level research projects. Her fashion art works have been selectd for many domestic and international art exhibitions, including art exhibitions in shanghai contemporary art museum and guangzhou contemporary art museum, beijing international fashion art exhibitions, and fashion art exhibitions in korea, germany, netherlands and italy. Her fashion art works have been nominated for awards in national art exhibitions for many times and have won many awards in zhejiang art exhitions.

谢梦狄
北京

清华大学美术学院服装艺术设计专业获学士学
位，随后赴英国伦敦时装学院学习时尚管理。目
前任教于中央美术学院设计学院时装专业。
2019年，作品《一屏江水》入选第十三届全国
美展，并被中国丝绸博物馆收藏；参加"千年
对话"时装艺术国际特邀展设计北京"经典映
像——时尚特邀展"。在清华大学"服装设计与
可持续发展国际论坛"作主题演讲。2018年作
品参展法国巴黎"中国设计四十年——传承与创
新"展，并在"中法设计论坛"作主题演讲；参
与法国巴黎 老佛爷百货中国新年橱窗设计

熊艺
贵阳

中央美术学院硕士研究生，贵州文化旅游职业学
院教师，贵州商学院外聘教师；
2018年，作品《织植纸制》入选首届"纺织之
光"中国纺织艺术展获"优秀奖"获"红衣坊
杯"世界华服时装设计大赛"最佳创意奖"；
2017年，作品《雀》受邀参加"韩2017年韩国
国际时装双年秀"作品《东方故事》获中国北京
大学生造型设计大赛三等奖；
2016年，作品《忆》获"中国风"全球男装设
计"最具商业潜力奖"作品《忆·构》获武汉时
装周"红T杯"优秀奖；
作品《丝路》入选2016年韩国第十六届FABI国
际交流时装展，受邀赴韩参展 / 韩国.首尔(韩国
崇实大学)；
作品《东方故事》赴英参加英国曼切斯特艺术学
院联合展演。

徐秋宜
中国台湾

毕业于法国巴黎高等装饰艺术学院(ENSAD)服
装设计系硕士，获第八届年轻服装设计师之夜 –
首奖，法国年轻设计师大赛 Beautiful Textil –
首奖、欧洲国际年轻服装设计师比赛金线首奖，
随后自创服装品牌本真一衣 NAIF by hsu chiu。
在法国完成服装设计硕士学位后，又取得台湾艺
术大学创意产业所博士学位，是台湾少数拥有艺
术创作、设计创意与文化管理的背景的创作者。
透过不断实验与研发，以及大胆叛逆的实验精
神，徐秋宜专注服装材料与版型的设计研究，为
时装衍生及创造文化价值，除了是服装设计师，
亦是跨领域行动艺术家，目前专职艺术创作，并
担任亚东技术学院材料与纤维系副教授。

Xie Mengdi
Beijing

Graduated from the Academy of Arts &
Design, Tsinghua University with a B.A. in
Fashion and Textile design, she went to
London collage of Fashion to learn fashion
management and obtained a M.A. degree
with distinction. At present, she teaches
fashion in the School of Design, China
Central Academic of Fine Arts.
In 2019, Fashion work Growing River was
selected into the 13th National Exhibition of
Fine Arts China, and collected by China Silk
Museum.
Attended "Millennium Dialogue" International
Fashion Art Exhibition. Attended
REFLECTIONS ON CLASSICS-Invitational
Exhibition on Fashion" Attended. 2019
International Forum on Fashion Design
and Sustainable Development, Tsinghua
University, Beijing, China
In 2018, "Attended Forty Years of Chinese
Design-Inheritance and Innovation"
exhibition in Paris, France, Keynote speech
of "Sino-French Design Forum". Chinese
New Year window design for Galeries
Lafayette in Paris, France

Xiong Yi
GuiYang

Master of Central Academy of Fine Arts,
Teacher of Guizhou Vocational College of
culture and tourism.
External teacher of Guizhou business school.
The work was selected into the first "light of
textile" China Textile Art Exhibition and won
the "Excellence Award" / China. Beijing / 2018
Weaving and planting paper won the "Best
Creativity Award" in the "hongyifang Cup"
World Chinese fashion design competition /
Nanchang, China / 2018
The work Bird was invited to participate
in the "2017 Korea International Fashion
Biennial Show" / Korea, Seoul / 2017
The work Oriental story won the third prize
of China Beijing college student modeling
design competition / China. Beijing / 2017
The work Memory won the "most commercial
potential award" of "Chinese style" global
men's wear design"/ China. Beijing / 2016
The work Memory. Structure won the "Red
T Cup" excellence award of Wuhan fashion
week / Wuhan, China / 2016
The work Silk Road was selected in the 16th
FABL international exchange fashion show
and invited to participate FABI / Seoul, South
Korea (Korea Chongshi University) / 2016
The oriental story participate in the joint
exhibition of Manchester Academy of art /
Manchester, UK / 2014

Hsu Chiu-i
Taiwan, China

Graduated with a master's degree in
Fashion Design from ENSAD, Paris,
France. She won the first prize of the 8th
Young Fashion Designer Night-First Prize,
the French Young Designers Exhibition
Beautiful Textil-First Prize, European
International Young Fashion Designer
Competition-Gold Thread First Prize,
followed by his own clothing brand "NAIF
by hsu chiu". After completing a master's
degree in fashion design in France, he
obtained a doctorate from the Creative
Industries Institute of Taiwan University
of the Arts who is a creator with a
background in cultural management.
Through continuous experimentation
and R&D, and a bold and rebellious
experimental spirit, hsu chiu i focuses
on the design and research of clothing
materials and patterns to derive and
create cultural value for fashion. In addition
to being a fashion designer, he is also a
cross-field action artist. He is currently
full-time artistic creation, and served as
an associate professor in the Department
of Materials and Fibers, Asia Eastern
University of science and technology.

闫洪瑛
北京

北京工业大学艺术设计学院服装与服饰设计专业副系主任，副教授。从事教学工作已有18年，研究领域：无障碍服装结构与技术研究 、传统服饰文化发展与创新设计研究、时装定制设计与技术研究。她主持、参与多项科研项目研究，并先后在主要期刊发表论文12余篇，作品参展12项，其中国际展8项，省部级展4项；作品发表10项，其中核心期刊1项。个人获奖7项，其中省部级及以上奖项4项。2009年，获得首都国庆60周年群众游行方阵策划设计"优秀工作者"称号。2014年，设计作品《清平盛世》参加2014年亚太经合组织（APEC）会议领导人服装设计，获"突出贡献纪念奖"。

严宜舒
中国香港

博士，服装与纺织品创新设计、数码针织设计实践与研究者。分别于香港理工大学、伦敦艺术大学及南京艺术学院获得博士、硕士及学士学位。作品曾在美国、英国、法国、日本、西班牙、墨西哥、孟加拉等多个国家及地区展出并获奖。2019年作品被中国丝绸博物馆收藏，2020年在香港理工大学服装馆举行个人作品展。

杨大伟
中国香港

英国诺森比亚大学时装专业毕业后从事设计工作，1997后从事设计教育工作，成为香港时装设计项目负责人。2009年，获得设计管理硕士学位并在英国任教。在土耳其伊兹密尔任教，并为美国学生进行在线教学。自2014年以来一直在中国任教。是北京服装学院国际艺术与设计课程的课程负责人。2016—2018年，任中央美术学院国际工作室主任。目前受聘于英国南安普顿大学，在大连教授学生时装设计。此外，参加了在美国、韩国和中国的许多学术演讲和展览。于2020—2021年被任命为美国林登伍德大学时尚商业与设计顾问委员会成员，并于2021—2025年为中国时装艺术国际同盟（IFAN）成员。

Yan Hongying
Beijing

Deputy dean of fashion and costume design department, School of art and design, Beijing University of Technology. Associate professor. She presided and participated in a number of scientific research projects.She has published more than 12 papers in major journals; 12 works art exhibition, including 8 International Exhibition, 4 Provincial and ministerial level exhibition, 2 Excellent Awards, 1 Memorial Awards; 10 works published, one published in the core journal. She has won 7 awards, including 4 provincial and ministerial awards and above. "Capital Excellent Worker", Planning and design of the square array of the mass March, The 60th Anniversary of The National Day,2009. "Outstanding contribution Memorial Award", Design work *Flourishing age*, Clothing design for leaders of the 2014 APEC meeting,2014.

Yan Yishu
Hongkong, China

PhD, practical researcher of innovative fashion and textile design and digital knitting design. She obtained her doctor's, master's, and bachelor's degrees from The Hongkong Polytechnic University, University of Arts London, and Nanjing University of Arts, respectively. Her art and design works have been exhibited and awarded in the USA, UK, France, Japan, Spain, Mexico, and Bangladesh. Her fashion design work has been collected by China National Silk Museum in 2019. A solo exhibition of Yishu's design research has been held in Fashion Gallery, The Hongkong Polytechnic University in 2020.

Yeung David
Hongkong, China

After graduation from Northumbria University, UK majoring in fashion, he worked as a design practitioner and after he worked in design education and had become a program leader in fashion design in Hongkong. In 2009, he achieved an MA in Design Management and taught in UK. He also taught in Izmir and did online teaching for USA students. He has then taught in China since 2014. He was Course Leader in an international art and design program at Beijing Institute of Fashion Technology. Between 2016 – 2018, he was Head of Studio at China Central Academy of Fine Arts. Currently, he has been hired by University of Southamption, UK to teach students fashion in Dalian.

David has also participated in many scholarly presentations and exhibitions in US, Korea and China. And lately, he has been appointed as Advisory Council Member by Fashion Business and Design, Lindenwood University, USA between 2020-2021, and Member of International Fashion Art Network (IFAN), China between 2021-2025.

杨秋华
深圳

1999年毕业于上海戏剧学院舞台美术系，现任深圳职业技术学院艺术设计学院服装与服饰设计专业副教授、二级舞美设计师，中国流行色协会委员，中国舞台美术协会会员。从事服装设计、纤维艺术等的教学、创作和研究。时装画作品曾获得首届全国高校教师时装画大赛银奖，入选第三届中国时装画大展；设计作品曾获第二届全国戏剧文化奖、第十一届广东省艺术节舞台美术奖，首届大湾区高校美术作品大赛二等奖等大小奖项三十多项，并多次入围和参加国内外重要展览，如上海国际纤维艺术展、国际防染艺术联展、中国现代手工艺学院展等。

于一鸣
上海

现就读于上海商学院服装与服饰设计专业本科，曾获第五届汇创青春工艺美术类一等奖。

余一萌
北京

中央美术学院时装方向教师，专业召集人。主要研究智能材料、可穿戴设备及未来时尚商业模式。英国皇家艺术学院时尚设计硕士，中央美术学院艺术与设计学士，曾赴巴黎高等装饰艺术学校交流学习。作品曾展出与展演伦敦时装周，中国国际时装周，故宫博物院与太庙等，报道与出版于*WWD, Vogue Italia, Sportswear International Magazine*，《服装设计师》等。

Yang Qiuhua
Shenzhen

Graduated from the stage art department of Shanghai Academy of drama in 1999, is now an associate professor of fashion and costume design in the Art Design College of Shenzhen Vocational and technical college, a second-class choreographer, a member of China Fashion Color Association and a member of China Stage Art Association. Engaged in teaching, creation and research of fashion design and fiber art. Fashion paintings won the silver prize in the first national college teachers' fashion painting competition and were selected into the third China Fashion painting exhibition; The design works have won more than 30 large and small awards, such as the second National Drama Culture Award, the stage art award of the eleventh Guangdong Art Festival, the second prize of the first Dawan District college art competition, and have been shortlisted and participated in important exhibitions at home and abroad for many times, such as Shanghai international fiber art exhibition, international anti dyeing art joint exhibition, China Institute of modern crafts exhibition, etc.

Yu Yiming
Shanghai

Studying in a bachelor's degree majoring in clothing and clothing design at Shanghai Business School. She has won the first prize in the fifth Huichuang Youth Arts and Crafts category.

Yu Yimeng
Beijing

Lecturer of Central Academy of Fine Arts, School of Design, Director of Fashion Design. Specializing in smart textile, wearable technology and future fashion business model. She graduated from Royal College of Art MA Fashion Womenswear with Distinction and hold the Bachelor's degree in Art and Design from Central Academy of Fine Arts, also had exchange study at ENSAD Paris. Her works have been exhibited and performed at London Fashion Week, China International Fashion Week, The Palace Museum, The Imperial Temple, etc.; published and launched in *WWD, Vogue Italia, Sportswear International Magazine*, *Fashion Designers*, etc.

庾晨溪
北京

本科毕业于天津美术学院 服装与服饰设计；
现就读于中央美术学院时装艺术。

袁大鹏
武汉

2006年获第六届中国服装设计最高奖评比一等奖；
2004年获第五届国际女装设计大赛银奖；
1996年获中国青年时装设计大赛金奖；
2008年获北京奥组委文化活动部授予纪念奖；
2009年第十一届全国美展 入选作品,第九届全国高校美术作品学术奖；
2014年第十二届全国美展获奖提名（APCE）国家领导人服装设计贡献奖；
2018年中国纺织艺术展 银奖；
2019全国中青年艺术（书画）成果展；
2019第十三届全国美展艺术设计展进京作品。

袁燕
厦门

1981年生于山东青州，现居中国厦门；
现任福州大学厦门工艺美术学院服装系副主任，副教授，硕士生导师；现为中央美术学院访问学者；
2018年福建石狮《系系之水》获福建省时尚设计大奖，本人并获2018年度福建省优秀服装设计师称号；
2019年中国武汉《无序之序》受邀参加首届国际可穿戴艺术展；
2020年中国厦门《不忘 · 海》参加并联 · 在场与介入当代艺术展。

Yu Chenxi
Beijing

Graduated from Tianjin Academy of Fine Arts
Now studying in Central Academy of Fine Arts

Yuan Dapeng
Wuhan

He won the first prize of the Sixth China fashion design highest award competition, In 2006, and the silver award of the Fifth International Women's fashion design competition in 2004. The gold award of China Youth Fashion Design Competition in 1996.
He was awarded the Commemorative Award by the cultural activities Department of 2008 BOCOG, In 2008, Won the 9th National Academic Award for art works in Colleges and universities in 2009 selected works in the 11th National Art Exhibition in 2009, and nominated for awards in the 12th National Art Exhibition and the national leader fashion design contribution award (APCE) in 2014
Silver Award of 2018 China Textile Art Exhibition
2019: national young and middle-aged Art (calligraphy and painting) achievement exhibition
13th National Art Exhibition Art Design exhibition works in Beijing

Yuan Yan
Xiamen

Born in Qingzhou, Shandong Province in 1981 ,and now lives in Xiamen, China.
At present hold the deputy director, associate professor and master supervisor of the Department of clothing of Xiamen Academy of Arts and crafts, Fuzhou University. I am now a visiting scholar of the Central Academy of fine arts.
2018 *Water of silk* won the fashion design award of Fujian Province, and I won the title of excellent fashion designer of Fujian Province in 2018, Shishi Fujian.
2019 *The order of disorder* was invited to participate in the first international wearable art exhibition, Wuhan.
2020 *Never forget the sea* participated in the parallel, presence and involvement contemporary art exhibition,Xiamen.

苑国祥
上海

博士，东华大学服装与艺术学院副教授，诺维萨德大学客座教授，中国文物学会纺织专业委员会会员。参与50多个国际服装和纺织品设计展，作品在中、日、泰、美、英、西班牙等多个国家和地区展出。

曾凤飞
厦门

中国时装设计最高奖"金顶奖"获得者；
全国纺织业劳动模范；
2014APEC领导人"新中装"主创设计师之一；
2017年厦门金砖五国峰会礼品及接待人员服装设计师；
先后六次荣获"中国最佳男装设计师"称号；
中国服装设计师协会时装艺术委员会主任委员；
亚洲时尚联合会中国委员会委员；
福建省服装设计师协会荣誉会长；
厦门服装设计师协会会长。

张刚
株洲

2001.07至今湖南工业大学（教师）；
2001毕业于山东工艺美术学院本科；
2010毕业于东华大学硕士；
2014.09—2015.06中央美术学院访学；
2015参加"间"—2015中韩时装艺术交流展；
2015参加"绽放"—2015时装艺术国际展；
2016参加"蓝之韵"—2016时装艺术国际展；
2017参加2017北京国际设计周"纸之为物"纸生活设计展；
2018参加"丝绸故事"2018北京时装艺术国际展；
2018参加2018首届"纺织之光"中国纺织艺术展；
2019参加2019"千年对话"2019时装艺术国际特邀展。

Yuan Guoxiang
Shanghai

Ph.D, Associate Professor of College of Fashion and Design, Donghua University. He is also a Visiting Professor in Novi Sad University and is a member of textile committee in China Cultural Relics Academy. His fashion and textile designs have been exhibited in over 50 international exhibitions in China, japan, Thailand, USA, UK, Spain and so on.

Zeng Fengfei
Xiamen

The Golden Award of China Fashin Association
The National textile industry model worker
2014APEC leader "new Chinese costume" main creative designer
2017XIAMEN BRICS summit gifts and reception staff costume designers
Six times of Top Chinese Menswear Designers
Committee member of The fashion arts council OF China Fashion Designer association
Member of the China committee of the Asian fashion federation
Honorary president of Fujian province Fashion designer association
President of XiaMen Fashion designer association

Zhang Gang
Zhuzhou

Hunan University of Technology / Lecturer
2001: graduated fron shandong university of art & design / Undergraduate
2010: graduated from donghua university/ Master.
2014-2015: annual visiting scholar of central academy of fine arts
2015: particited China and Korea fashion arts exhibition *Beteen*
Particited the *Bloom*-2015 International Fashion Art Exhibition
2016: *BLUE*-2016 International Fashion Art Exhibition (China-Ordos)
2017: Beijing international design week *By material of paper* paper design exhibition of life
2018: Has particited the "Silk story-2018 International Fashion Art Exhibition"
2018: Particited The First "Light of Textile" China Textile Art Exhibition

张国云
重庆

毕业于清华大学美术学院。近年来致力于传统服饰文化传承与创新研究，主持国家、省部级、校级等项目。出版著作1部，发表核心期刊论文20余篇。作品入选第十三届全国美展，第七届北京国际美术双年展等。

张鹏
北京

毕业于清华大学美术服装艺术设计系，现任北京工业大学艺术设计学院服装与服饰设计系系主任、副教授、硕士生导师。主持并承担北京市社会科学基金项目、北京市教委社科计划项目及专业教育教学研究课题十余项。时装艺术作品入选Fashion Art时装艺术国际展（2013—2019年）、国际时尚艺术与设计展、韩国"Fashion Meets Jewerly"时装艺术展、日本福冈中国·韩国·日本当代艺术展、日本宫城中日艺术与手工艺作品交流展、2019年度时尚回顾展等多项国内外展览，并被中国丝绸博物馆收藏。

张清心
海口

海南师范大学美术学院服装与服饰设计专业教师；香港理工大学硕士；知璞品牌创始人。多年来致力于民族织锦的传承与发展研究，进行服饰产品的设计与研发。服装设计作品获得2012中国国际服装创意设计大赛最佳创意奖，作品被中国丝绸博物馆收藏。

Zhang Guoyun
ChongQing

I graduated from the Academy of fine arts of Tsinghua University. In recent years, I has been committed to the research on the inheritance and innovation of traditional clothing culture, and presided over national, provincial and school level projects. I had published one book and more than 20 papers in core journals. My works were selected in the 13th National Art Exhibition and the 7th Beijing International Art Biennale.

Zhang Peng
Beijing

I graduated from the Department of Fashion Design, Academy of Arts & Design, Tsinghua University. I work as department head, associate professor and master's supervisor in the Department of Fashion Design, College of Art and Design, Beijing University of Technology. I preside over and undertake more than 10 projects, such as Beijing Social Sciences Fund project, social science project of Beijing Municipal Education Commission, research project of professional education and teaching, etc.My fashion art works have been selected into many domestic and foreign exhibitions, such as Fashion Art International Exhibition (2013–2019), International Fashion Art and Design Exhibition, South Korea "Fashion Meets Jewerly" Fashion Art Exhibition, China+Korea+Japan Contemporary Art Exhibition, Fukuoka, Japan, Sino-Japanese Art and Handicraft Works Exchange Exhibition, Miyagi, Japan, 2019 Fashion Review, etc., and have been collected by China National Silk Museum.

Zhang Qingxin
Haikou

I am a teacher majoring in fashion and costume design in the school of fine arts of Hainan Normal University. I graduated from Hongkong Polytechnic University with a master of Arts degree.I have been committed to the research on the inheritance and development of national brocade for many years. My fashion design works won the best Creativity Award in 2012 China international fashion creative design competition, and my works were collected by China Silk Museum.

张婷婷
北京

中华女子学院副教授，中国女性文化遗产研究中心主任。近十年来个人作品广泛陈列展示于北京故宫太庙艺术馆、北京798-A12T当代艺术中心、北京爱慕美术馆、中央美术学院美术馆、北京筑中美术馆、今日美术馆、北京炎黄艺术馆、中国丝绸博物馆、苏州丝绸博物馆、重庆大足石刻博物馆河南美术馆等。多系列作品随国家级文化艺术巡展至美国、土耳其、泰国、法国、希腊、埃及、俄罗斯、瑞典等。受邀北京国际艺术品基金会、北京艺术沙龙、德国HAFELE五金集团、纽约州立大学、纽约州首府奥尔巴尼Pine Hill图书馆、中美友协等机构，举办《卷香丝语》《缝合》《墨》《花好月圆》《花样年华》《温暖技艺》《卷香丝语之千里江山》等系列个人展览及工作坊。作品先后被中国丝绸博物馆、苏州丝绸博物馆以及法国、德国、日本等国机构和个人永久收藏。

郑程元
北京

中央美术学院时装艺术方向研究生

郑贤
韩国

2021年参加时装艺术国际展（济州：济州国际和平中心）；
2020年参加秋季KOSCO展览会（尚州：韩国韩博促进研究所）；
2019年参加"小型婚礼礼服"展（首尔：Mubongheun）；
2019年参加"和平与分享"平昌冬奥会一周年庆典。

Zhang Tingting
Beijing

Associate professor of Art College of China Women's University, Director of Chinese Women's Cultural Heritage Research Center, Deputy Director of Fashion Arts Committee of China Textile Engineering Society.Personal works widely display in nearly a decade in the ancestral temple of Beijing the imperial Palace Museum, Beijing 798-A12T contemporary art center, Beijing Love-art Museum, Zhuzhong academy of fine arts gallery, Beijing Yanhuang Art Gallery, Today Art Museum of Beijing, China silk Museum, Suzhou Silk Museum, Chongqing Dazu Grottoes, Henan Art Museum. Many series of works have been exhibited in the United States, Turkey, Thailand, France, Greece, Egypt, Russia, Sweden and other countries. Once by the Beijing international art foundation, Beijing art salon, Germany HAFELE hardware group, New York State University, PineHill Library in Albany, NY, and China-US Friendship Association (New York). A series of personal exhibitions and workshops were held, such as *Silk Language of Scrolls*, *Stitching*, *Ink*, *Flowers and Moons and Warm Skills* Exhibitions and workshops Many of her works have been permanently collected by Chinese and foreign institutions such as The China Silk Museum and Suzhou Silk Museum.

Zheng Chengyuan
Bcijing

Central Academy of Fine Arts Fashion art research direction

Jung Hyun
Korea

2021 Fashion Art International Exhibition (Jeju: International Peace Center Jeju)
2020 Fall KOSCO Exhibition (Sangju: Korea Hanbok Promotion Institute)
2019 "Dress for Small Wedding" (Seoul: Mubongheun)
2019 "Peace & Share", 1st Anniversary Festival of PyeongChang Olympic Winter Games

周朝晖
长沙

硕士，教授、一级舞美设计师、中央美院访问学者、湖南省政府首批"三百文艺人才"入选专家、时装艺术国际同盟常务理事、中国纺织工程学会时装艺术专业委员会常务副主任、湖南省设计艺术家协会副秘书长；湖南省舞台美术学会常务理事；
1. 完成了国家艺术基金项目《湖南少数民族织锦文化创意与工艺创新人才培养》等多个国家级、省级项目；主编教材多本。
2. 2018年举办了"曼舞霓裳——周朝晖、包小青舞蹈服饰设计展演"；2020年举办了"苗衣绣艺——苗族服饰艺术展"；
3. 有《自由飞翔的蝶》《梵花盛开》《绝恋》《怜相伴》《古窑陶裳》等8个作品入选时装艺术国际展；
4. 担任原创音乐剧《同一个月亮》、话剧《永不凋谢的姊妹花》，湘剧《楚辞》《忠诚之路》等多个大型剧目、晚会服装设计，并有多个剧目获奖。

周梦
北京

中央民族大学教授；
博士研究生导师；
2005年硕士毕业于北京服装学院；
2005年至今于中央民族大学美术学院任教；
两项国家社科基金项目主持人；
设计作品入选全国美展。

周璇
北京

2014年9月至2019年6月就读于中央美术学院造型学院雕塑系公共艺术工作室，同年考入中央美术学院雕塑系研究生；
2019年作品《衣旧出彩》参展景德镇陶溪川美术馆"贴地飞行展览"；
2020年798国际艺术交流中心"生声——中国当代雕塑作品展"担任策展人；
2020年作品《苹果是苹果的果实》参加中央美术学院美术馆"北京青年艺术双年展"；
2021年新国贸饭店"我就粉你——粉色当代先锋艺术展"担任策展人；
2021年西单更新场"恋与喵星人公益展"担任策展人；
2021年新国贸饭店"世界是银色的——银色当代先锋艺术展"担任策展人。

Zhou Zhaohui
Changsha

Master degree, professor, and first class stage designer, visiting scholar of Central Academy of Fine Arts, selected experts of Hunan provincial government's first batch of "300 literary talents", executive director of Fashion Art International Alliance, executive deputy director of Fashion Art Professional Committee of China Textile Engineering Society, deputy secretary general of Hunan Design Artists Association, executive director of Hunan Institute of Stage Design.
1.Accomplished the National Art Fund project- Cultivation of Creative and Technological Innovation Talents for the Brocade Culture of Hunan Minority Nationalities, and multiple national and provincial projects, and edited several textbooks.
2.In 2018, held an exhibition called—The Gorgeous Dancing Dress- Dance Costume Design Exhibition of Zhou Zhaohui and Bao Xiaoqing. In 2020, held an exhibition called- The Art of embroidery of Miao costume-
Miao Costume Art Exhibition.
3.Eight works were selected for the International Exhibition of Fashion Art. For example: *Free flying butterfly, the flower of Buddhism in full bloom, Beautiful love, Dependence and Accompany, Ancient Ceramic clothes and so on.*
4. As a costume design in many Large-scale repertoires: *an original musical called The Same Moon, a drama called The Undying Sisters,* Hunan operas called *The Songs of Chu* and *the road of loyalty.* In addition, many plays have won awards.

Zhou Meng
Beijing

Professor of MUC, doctoral supervisor. Got master's degree from Beijing Institute of Fashion Technologe in 2005 and has been on the faculty in MUC since 2005.
Host of two National Social Science Foundation Projects, and design works were selected into the national art exhibition.

Zhou Xuan
Beijing

From September 2014 to June 2019, she studied in the public art studio of the Sculpture Department of the school of plastic arts of the Central Academy of fine arts, and was admitted to the graduate student of the Sculpture Department of the Central Academy of Fine Arts in the same year.
In 2019, the work *old clothes shine* was exhibited in the "ground flight Exhibition" of taoxichuan Art Museum in Jingdezhen.
In 2020, she served as the curator of "*Shengsheng-Chinese Contemporary Sculpture Exhibition*" of 798 international art exchange center.
In 2020, the work *apple is the fruit of apple* participated in the "Beijing Youth Art Biennale" of the Art Museum of the Central Academy of fine arts.
In 2021, she acted as the curator of "*I'll pink you-Pink contemporary pioneer art exhibition*" in New International Trade Hotel.
In 2021, she acted as the curator of the "*love and meow star people public welfare Exhibition*" in Xidan update field.
In 2021, she acted as the curator of "*the world is silver-Silver contemporary pioneer art exhibition*" in New International Trade Hotel.

卓克难
重庆

四川美术学院设计学院服装设计系副教授；
1982 年毕业于四川美术学院工艺系染织本科专业；
曾任广州罗曼达制衣有限公司设计总监；
多件作品在《中国服装》及其他专业画册上发表，
还多次为企业或文化活动进行服装项目的设计；
水彩画作品获"第二届重庆市水彩、水粉画展览"
优秀奖，"第三届重庆市水彩、水粉画展览"优秀奖；
2013 年时装画获"首届金苑杯中国时装画大展"
优秀奖，第三届"鲁绣杯"中国大学生家用纺织
品创意大赛优秀指导教师奖。连续三届（2013、
2014、2015）获得中国国际面料创意大赛优秀指
导教师奖；
2015 年、2017 年二次获得中国服装设计师协会颁
发的优秀指导教师奖。

宗源
无锡

2020 年获东华大学服装与服饰设计学士学位；
2017 年获 BONOBOJEANS 牛仔服改造项目优
胜奖；
2018 年获第七届真维斯休闲装设计大赛东部三等
奖；
2019 年获第八届石狮杯全国高校毕业生服装设计
大赛优胜奖；
现为服装设计师及自由艺术家创作至今，是流连
于服装、插画、3d 虚拟装置的多领域创作者。

邹英姿
苏州

研究员级高级工艺美术师；中国工艺美术大师；中
国工美行业艺术大师；国家级非物质文化遗产传承
人（苏绣）；第十三届江苏省人民代表大会代表；
江苏省妇女联合会三八红旗手。
2008 年所独创的"邹氏滴滴针法"，是第一件获
得国家刺绣技术专利的刺绣技法，自 2011 年起，
获得具有自主知识产权的刺绣外观设计专利 92 项，
设计版权多项；"滴滴绣""英姿绣"和"英姿劈
针绣"为国家注册商标品牌。
作品多次获国内外展会大奖，作品《色空不二》被
中国国家博物馆收藏、作品《缠绕》被大英博物馆
收藏、作品《姑苏人家·醉花荫》被中国工艺美术
大师博物馆收藏；按照原比例复绣敦煌莫高窟流失
在英国大英博物馆的唐代刺绣《凉州瑞像图》捐赠
给敦煌研究院收藏。为国际友人和国家元首等名人
定制的刺绣肖像也深得各界好评。

Zhuo Kenan
Chongqing

Associate professor of Fashion and Textile
Design at Sichuan Fine Arts Institute. She
graduated from the Sichuan Fine Arts
Institute in 1982.
Used to be the design director of
Guangzhou Romanda Garment Co., Ltd.
Part of her fashion.
artworks have been published in *China
Fashion*, etc.
Kenan's watercolor works continuously
received the excellent award of the 2nd and
3rd Watercolor & Gouache Painting
Exhibition in Chongqing.
In 2013, Kenan award the excellent award
of the 1st Jinyuan Cup Chinese Fashion
Illustration Exhibition and the excellent
guide teacher award of the 3rd "Lu
Xiu"Chinese College Students' Home
Textile Design Competition.
She was the recipient of the China
International
Fabrics Design Competition for three
consecutive sessions (2013, 2014 and 2015).
In 2015 and 2017, she awarded the excellent
guide teacher by the China Fashion
Association for her patient guidance work.

Zong Yuan
Wuxi

Received a Bachelor of Fashion and
Accessory Design from Donghua
University in 2020.
Won the winning prize of BONOBOJEANS
denim clothing transformation project in
2017.
Won the third prize in the East of the
7th Jeanswest Casual Wear Design
Competition in 2018.
Won the 8th Shishi Cup National College
Graduate Fashion Design Competition in
2019.
Currently as a fashion designer and
freelance artist, a multi-field creator who
lingers in clothing, illustrations, and 3D
virtual installations to this days.

Zou Yingzi
Suzhou

Researcher-level senior craft artist; Master
of Chinese arts and crafts; Art masters in
China's industrial beauty industry; Inheritor
of national intangible cultural heritage
(Suzhou embroidery); Deputy to the 13th
Jiangsu Provincial People's Congress;
The March 8th red flag bearer of Jiangsu
women's Federation.
The "Zou's Didi Needlework" originally
created in 2008 is the first embroidery
technique to obtain the national
embroidery technology patent. Since 2011,
92 embroidery design patents have been
obtained with independent intellectual
property rights and many design copyrights.
"Didi Embroidery", "Yingzi Embroidery"
and "Yingzi Split Needle Embroidery" are
national registered trademark brands.
The works have won many awards in
exhibitions at home and abroad. The work
named *Color Is the Same* has been
collected by the National Museum of China, meanwhile,
the work named *Winding* has been collected
by the British Museum, moreover, *Gusu Family·
Drunken Flower Shade* has been collected by
the Chinese master of Arts and Crafts Museum.
According to the original proportion, the Tang
Dynasty embroidery *kyamuni preaching on the
Vulture Peak / Miraculous Image of Liangzhou*
lost in the Dunhuang Mogao Grottoes in the
British Museum was recreate and donated to
the Dunhuang Research Institute as collection.
The embroidered portraits customized for
international friends, heads of state and other
celebrities have also won high praise from all
walks of life.

邹莹
兰州

2008.9—2011.1 于北京服装学院攻读设计艺术
学硕士，"高级时装设计"方向；
2018.7—2018.8 参加国家艺术基金人才培养
《中国传统服饰图案传承与创新应用设计人才培
养》项目；
2018.9—2019.7 于清华大学美术学院以"染织
艺术设计、历史与理论研究"为题进行访问学者
工作；
2013.1至今于兰州城市学院艺术设计学院担任
专业讲师工作，任染织与软装工作室（服装工作
室）负责人。

区永欣
中国香港

时装造型师，以时装造型及撰写文章为主。曾任
职香港各大时装杂志，主要负责杂志内容编采、
时装造型硬照及封面拍摄，曾与各大时装品牌合
作，为影星及模特儿造型拍摄，合作对象包括汤
唯、桂纶镁、Jessica（少女时代成员）、Alexa
Chung、林忆莲及中田英寿等。

张丽蓉
中国香港

独立艺术家。自2013年于毕业皇家墨尔本理工
大学获得了美术学士学位以来，以自由职业者的
身份支持她作为艺术家的全职艺术实践。
灵感大部分来自对自然的敏锐观察以及对上帝创
造的好奇心。加上与人，空间，形式和文化的互
动，研究各种传达人类潜意识状态的视觉语言。
曾于香港及不同国家举办个人展览、表演与联展。

Zou Ying
Lanzhou

2008.9–2011.1 Master of Arts and Design
at Beijing Institute of Fashion Technology,
direction "Fashion Design".
2018.7–2018.8 Participated in the National
Art Fund Talent Training Project "Chinese
Traditional Costume Pattern Inheritance
and Innovative Application Design Talent
Training".
2018.9–2019.7 Visiting scholar work with
the topic of "Dyeing and Weaving Art
Design, History and Theoretical Research"
at the Academy of Fine Arts of Tsinghua
University.
2013.1–now Work as a professional
lecturer in the School of Art and Design of
Lanzhou City University, as the head of the
dyeing, weaving and soft clothing studio
(clothing studio).

Kim Au
Hongkong, China

Experienced fashion editor and stylist,
now working as a freelancer, Kim had
worked for various fashion magazines, she
styled and art directed fashion editorials
and covers, as well as planning and writing
fashion features. Over the years, Kim had
styled various celebrities, like Tang Wei,
Kwai Lun Mei, former "Girls Generation"
member Jessica, Alexa Chung, Sandy Lam
and Hidetoshi Nakata to name a few.

Lio Cheung
Hongkong, China

Independent artist based in Hongkong.
Since graduation with a bachelor's degree
in Fine Arts from RMIT University 2013,
She works as a freelancer to support her
full-time art practise as an artist. Most
of her inspiration come from her keen
observation of nature and her curiosity
about God's creation. Coupled with
interactions with people, space, form and
culture, she is interested in experimenting
with visual languages that conveys human
subconscious states.
She has joined different exhibitions in
Hongkong and different countries.

主办单位
Organizer

国际合作
International Cooperation Association

# 中国·西樵
## XIQIAO, CHINA

## 文翰樵山·最岭南
### Extraordinary Xi Qiaoshan Mountain
### Distinctive Lingnan Culture

西樵镇位于佛山市南海区西南部，地处珠江三角洲腹地，是中国面料名镇、中国龙狮名镇、国家卫生镇、国家特色小镇、国家历史文化名镇、广东省教育强镇、广东省中心镇、广东省旅游名镇、广东省森林小镇。全镇面积 176.63 平方公里，是全国农村综合性改革试点试验地区之一。

Xiqiao Town, located in the southwest of Nanhai District, Foshan City,Guangdong province, China. and in the center of the Pearl River Delta, is a famous Chinese fabric town, a famous Chinese dragon and lion town, a national health. town, a national Characteristic Small town, a famous national historical and cultural town, a strong town of education town in Guangdong Province, a central town in Guangdong Province, a famous tourist town in Guangdong Province, and a forest town in Guangdong Province. It covers an area of 176.63 square kilometers and is one of the pilot areas of China's comprehensive rural reform.

西樵自然资源丰富，辖区内有"珠江文明的灯塔""南粤名山"之美誉的西樵山风景名胜区，是国家重点风景名胜区、国家 AAAAA 级旅游景区、国家森林公园、国家地质公园，听音湖公园获授"广东省宜居环境范例奖"，位居"佛山十大最美公园"榜首。山南片区基塘农业系统入选中国重要农业文化遗产名单。中国文艺青年小镇与自然教育营地项目落户西岸片区。以吉赞横基、民乐窦、吉水窦为重要组成部分的桑园围入选世界灌溉工程遗产名录。

Xiqiao is rich in natural resources with the famous Xiqiao Mountain Scenic Spot located in this town, which is well-known as the Lighthouse of Pearl River Civilization and Famous Mountain of South Guangdong. Now this scenic spot is a national key scenic spot, a national 5A-level tourist attraction, a national forest park and a national geopark, in which the Tingyin Lake has been awarded the "Guangdong Province Habitat Environment Example Prize" and ranked at the top of "Foshan Top Ten Beautiful Parks". The dike-pond agricultural system in South Mountain District has been listed in China's important agricultural cultural heritage. In the West Coast District, the young Chinese artists town and the project of nature education camp has been constructed. The Sang Yuan Dikes, mainly composed of Jizan Hengji, Minle Dou and Jishui Dou, was successfully added into the list of World Irrigation Engineering Heritage.

西樵文化底蕴深厚，孕育了民族企业家陈启沅、清代尚书戴鸿慈、武术宗师黄飞鸿、中国妇女解放运动先驱区梦觉等一批著名人物，培育了国艺影视城、渔耕粤韵文化旅游园、听音湖观心小镇等精品旅游项目，打响了狮王争霸赛、听音湖龙舟赛、樵山文化节、翰林文化节等特色品牌。西樵醒狮多次登上央视舞台，西樵凭"醒狮"获"中国民间文化艺术之乡"称号。松塘村获评中国历史文化名村、全国乡村旅游重点名村，百西村头村、简村入选广东省历史文化名村。

With its deep cultural heritage, Xiqiao has nurtured a number of famous figures with such as the national entrepreneur Chen Qiyuan, high official in Qing Dynasty Dai Hongci, master of Kung Fu Huang Feihong, pioneer of Chinese women's liberation movement Qu Mengjue, etc., cultivated some good-quality tourism projects like the National Arts Studios, Cantonese Fishery and Agro-farming Culture Tourist Park, Guanxin Town by Tingyin Lake, etc., and created certain brands with its own features, including the Lion King Competition, Tingyin Lake Dragon Boat Race, Qiao Mountain Cultural Festival, Hanlin Cultural Festival, etc. Xiqiao Lion Dance has been on the CCTV for many times and been awarded as the Hometowns of Chinese Folk Culture and Art with its Lion Dance. Songtang village was recognized as a famous Chinese historical and cultural village and one of the national key villages for rural tourism; Baixicuntou village and Jian village was named to a famous historical and cultural village of Guangdong Province.

西樵产业发展蓬勃，形成以文旅、陶瓷、纺织、卫生用品、酒店饮食业为主的多元化发展产业体系。西樵纺织素有"广纱甲天下、丝绸誉神州"之美名，广东西樵纺织产业基地是"广东省十大循环产业基地"。蒙娜丽莎陶瓷被国家工信部列入首批绿色工厂名单，并获授"中国质量奖提名奖"。卫生用品产业加快集聚，西樵摘得"中国妇婴卫生用纺织品示范基地"国字招牌。现代农业根基稳固，何氏水产活鱼冷链运输技术和"何氏蹦蹦鱼"品牌日渐响亮。

Xiqiao industries are thriving and prosperous, forming a diversified industrial system with cultural tourism, ceramics, textiles, sanitary products, hotels and catering industry as the main drive. Xiqiao textiles have long been known as the "World's Most Famous Yarn and Silk", and the Xiqiao textile industry base in Guangdong is one of the Top Ten Recycling Industry Bases in Guangdong Province. Monalisa Tiles has been named to the list of first green factories by Ministry of Industry and Information Technology of people's Republic of China and been awarded a "China Quality Award (Nomination Award)". With the accelerated industrial agglomeration of sanitary products, Xiqiao has obtained a national title of "Chinese Textile Demonstration Base for Maternal and Infant Health". Modern agriculture of is also firmly rooted in Xiqiao, and He Shi live fish cold chain transport technology and its brand He Shi Beng Beng Yu is becoming more and more famous.

站在新的历史起点上，西樵大力推动一、二、三产业融合发展，进一步优化生产、生态、生活空间布局，奋力闯出高质量发展的西樵之路。

At a new historical starting point, Xiqiao is vigorously pushing forward the integrated development of primary, secondary and tertiary industries, further optimizing the spatial layout of production, ecology and living, and striving to create a path of high-quality development for Xiqiao.

# 美丽名镇·幸福家
## Beautiful Town-Happy Homeland

以西樵山为龙头，辐射听音湖片区、山南片区、西岸片区，持续汇聚优质文旅资源，加速发展全域旅游，重塑生态、生产、生活空间，打造新的产业增长点和动力源，擦亮"文翰樵山最岭南"文旅品牌，展现岭南文旅高地新气象。

With Xiqiao Mountain as the leader, radiating the Tingyin Lake District, South Mountain District and West Coast District, the town keens on continuous integration of high-quality cultural and tourism resources to accelerate tourism development in the whole area, reshape the ecology, production and living spaces, create new industrial growth points and driving forces, build up the cultural and tourism brand of "Extraordinary Qiaoshan with Distinctive Lingnan Culture" and spread out the new look of Lingnan cultural and tourism highlands.

### 听音湖片区 Tingyin Lake District

听音湖片区位于西樵山的西北山麓，规划总面积5.7平方公里。听音湖片区坐拥着西樵山国家5A级景区独特的自然、历史和人文资源，是国家旅游产业集聚（实验）区的重要载体，着力打造"岭南文旅RBD"，是展示岭南文化的重要窗口。目前，片区路网全面建成，樵山大道、锦湖大道、理学大道等多条道路完工通车，实现片区对外交通的快速联系。樵山文化中心、飞鸿馆全面投入运营，各类国际国内高端会议展览、大型文旅活动和体育赛事纷纷落户。观心小镇、希尔顿欢朋酒店相继开业，有为馆、听音湖粤菜美食集聚街区、听音湖夜游项目加快推进，云影琼楼、白云楼、樵顺嘉宝酒店提速建设，听音湖配套日臻成熟。特别值得一提的是，国内演艺龙头宋城集团投资的佛山千古情项目将开门迎客，预计每年为西樵带来100万人次以上的游客增量。

This district is located in the northwest foothills of Xiqiao Mountain, with a total planning area of 5.7 square kilometers. The national 5A scenic spot of Xiqiao Mountain with unique natural, historical and humanistic resources is located here, which makes this district an important carrier of national tourism industry agglomeration (experimental) zone to build a Lingnan Cultural Tourism RBD and an important window to show the image of Lingnan culture. At present, the road network is fully completed and a number of roads such as Qiaoshan avenue, Jinhu avenue and Lixue avenue, etc., have been completed and opened to traffic, with the whole district benefiting from such rapid connection to external traffic. Qiaoshan Cultural Center and Feihong Hall has been put into operation and various international and domestic conferences, exhibitions, large-scale cultural and tourism activities and sports events have been held here. Various supporting facilities in this district have been gradually perfected, with the Guanxin Town and Hampton by Hilton being opened for business, the Youwei Hall, Tingyin Lake Gourmet Street of Cantonese Cuisine and Tingyin Lake Night Tour Project being accelerated and the construction of hotles like Yunying Qionglou, Baiyun Lou and Qiaoshun Jiabao being speeded up. Particularly, the Foshan Romance Shows Project invested by the domestic leading performance company Songchegn Group will open for business and is expected to raise the number of visitors to Xiqiao by more than 1 million each year.

## 樵山文化中心 Qiaoshan Cultural Center

以樵山文化中心为载体，积极引进高端精品展会，大力发展会展产业和会议经济新业态。

Taking Qiaoshan Cultural Center as a carrier, Xiqiao will actively introduce high-quality exhibitions and vigorously develop the exhibition industry and the new business form of conference economy.

## 宋城·佛山千古情 Songcheng-Foshan Romance Shows

全国文旅龙头宋城演艺全力打造宋城·佛山千古情项目，聚合演艺娱乐资源，培育岭南文化演艺精品。

This project, created by the domestic leading performance company Songchegn Group, will integrate performance and entertainment resources and cultivate marvelous works with distinctive Lingnan culture.

## 有为馆 Youwei Hall

与广州美术学院合作建设广府美术馆，作为广府美术和岭南画派的艺术展陈基地，并将成为南海西部的公共文化中心。

Guangfu Art Gallery is scheduled to be set up together with Guangzhou Academy of Fine Arts and will serve as an art exhibition base for Guangfu Art and Lingnan Painting and as a public cultural center in western Nanhai.

## 飞鸿馆 Feihong Hall

定位为展览、表演、培训、赛事、各类文体协会基地和旅客体验、购物于一体的大型旅游武术文化综合馆，打造南中国武术中心。

This Hall is positioned as a large complex of tourism and Wushu culture, integrating exhibitions, performances, training, tournaments, various cultural and sports events and travelers' experience and shopping, to create the Wushu center in South China.

### 国艺影视城 National Arts Studios

加快"南方影视中心""国艺影视城"影视产业集聚发展核心区域建设，争创粤港澳大湾区影视产业合作试验区，打造世界级影视旅游创意产业基地。

The construction of Southern Film and Television Center and National Arts Studios, as the core of film and television industry concentrated development zone, has been accelerated, to create a pilot zone for film and television industry cooperation in Guangdong-Hongkong-Macao Greater Bay Area and a world-class creative industry base for movie and TV tourism.

### 西樵山民宿 Xiqiao Mountain B&B

大力发展乡村旅游和岭南特色民宿产业，打造西樵山上精品民宿标杆。

The town will vigorously develop bed and breakfast (B&B) industry with rural tourism and Lingnan characteristics and establish a benchmark of high-quality B&B on Xiqiao Mountain.

### 听音湖美食轩 Tingyin Lake Feast Village

首期建设 400 米滨水餐饮美食特色街区，并辅以酒吧等休闲、娱乐商业配套。

A 400-meter-long gourmet street by the lake and commercial facilities like bars and other leisure and entertainment businesses will be build up during the first phase of construction.

### 樵顺嘉宝酒店 Qiaoshun Jiabao Hotel

项目将建设高端精品酒店，完善听音湖片区的文旅配套，提升西樵山旅游接待能力。

A luxury boutique hotel will be constructed to improve the supporting facilities for Tingyin Lake District and to enhance the tourism reception capacity of Xiqiao Mountain.

### 西樵山登山索道 Xiqiao Mountain Ropeway

计划投资 1.2 亿元，对西樵山原有登山索道进行改造提升，进一步完善西樵山旅游配套设施。

It is planned to invest CNY 120 million to renovate and upgrade the original ropeway and further improve the tourism supporting facilities of Xiqiao Mountain.

### 中国文艺青年小镇与自然教育营地
### The Young Chinese Artists Town and Nature Education Camp

拟投资超 50 亿元，建设生态主题公园、高端精品民宿、乡村文旅产业共享办公空间、文化艺术空间、自然教育空间、自然运动空间、自然教育营地、高端餐饮与商业服务配套等。

It invests over CNY 5 billion to build ecological theme parks, high-end boutique B&Bs, shared office space for rural cultural tourism industry, cultural art space, nature education space, nature sports space, nature education camps, high-end catering and other commercial services, etc.

### 听音湖龙舟赛 Tingyin Lake Dragon Boat Race

深化"旅游 +"融合发展模式，积极举办听音湖龙舟赛等各项影响力的体育竞技赛事、文化活动。

Various influential sports and cultural activities like Tingyin Lake Dragon Boat Race, etc., are frequently held to deepen the "tourism +" integrated development model.

### 云影琼楼及白云楼 Yunying Qionglou and Baiyun Lou

云影琼楼高端休闲度假酒店——云影琼楼与白云楼整体改造项目，打造成为精品度假酒店

The two hotels will be integrally converted into a boutique resort hotel-Yunying Qionglou high-end relax resort.

## 观心小镇 Guanxin Town

融合岭南文化、精品餐饮、生态旅游、度假休闲、商务会议于一体、涵盖"吃喝玩乐住"丰富业态的文旅休闲旅游综合体。

Guanxin Town It is a cultural and leisure tourism complex, integrating Lingnan culture, fine dining, ecological tourism, relaxed vacation and business meeting, and covering a variety of business modes of entertainment like eating, drinking, playing and living, etc.

## 环西樵山水上生态观光线
## Sightseeing Route of Aquatic Ecosystem Around Xiqiao Mountain

加强"一河两岸"绿化景观建设，在小岛、村落、山岗建设水上码头和景观楼台，重塑岭南水乡风貌，打造从山南至听音湖 10 公里水上生态观光线路。

In order to enhance the construction of "One River-Two Banks" green landscape, water docks and landscape terraces have been build up in small islands, villages and on the hillocks, to reshape the landscape of water country with Lingnan features and styles and to create a 10-kilometer water ecological tourism route from South Mountain to Tingyin Lake.

# 蓝图 / 未来规划
## Blueprint / Future Planning

## 空间再造，产业重塑，缔造美好生活在西樵
### Creating a Better Life in Xiqiao with Recycled Space and Restructured Industries

西樵镇将以"空间再造，产业重塑，缔造美好生活在西樵"为工作主线，文旅产业、传统产业和新兴产业并驾齐驱，构建多元化现代产业体系。

Xiqiao Town takes "Creating a Better Life in Xiqiao with Recycled Space and Restructured Industries" as the guideline, with cultural and tourism industries, traditional industries and emerging industries as the driving forces to build a diversified modern industrial system.

改革开放 40 年来，村级工业园作为西樵镇产业经济发展的细胞，是承载农村经济发展的重要载体，为西樵镇的经济腾飞奠定了基础．随着城市升级和产业转型持续推进，提升改造村级工业园成为拓展城市发展空间、优化产业布局、改善人居环境的主攻方向。西樵镇村级工业园改造提升工作全面铺开，规划五大主题产业片区，总面积约 14115 亩，全力构建西樵产业空间新格局。

Over the past 40 years of China's reform and opening up, village-level industrial park, as a cell of the industrial and economic development of Xiqiao, has become an important carrier of rural economic development and has laid a solid foundation for rapid economic growth of Xiqiao. With the continuous development of urban upgrading and industrial transformation, the upgrading and transformation of village-level industrial parks has become the main direction of expanding urban development space, optimizing industrial layout and improving the living environment. Xiqiao has conducted an overall transformation and upgrading work for its village-level industrial parks, with five thematic industrial districts and a total area of about 9.41 million square meters planned for this programme, to build a new pattern of industrial development in Xiqiao.

### 山北片区：新材料 + 智能装备制造社区
### North Mountain District : new materials + intelligent equipment manufacturing

片区未来产业定位是新材料与智能装备制造业，以通用设备制造业、汽车配件制造业、电气机械器材制造业和新材料等四大产业为产业主体，着力将山北片区打造成为西樵镇的高端制造产业社区。

The future industry orientation of this district will focus on new materials and intelligent equipment manufacturing, with four major industries such as general equipment manufacturing, auto parts manufacturing, electrical and mechanical equipment manufacturing and new materials, etc., to build this district into a high-end manufacturing industry community in Xiqiao.

### 听音湖拓展区：文旅 + 现代商贸
### Tingyin Lake Extension District : cultural tourism + modern commerce and trade

听音湖片区是打造"文翰樵山"岭南文化旅游高地的核心，拓展区作为听音湖片区的延伸，着重引进文化创意、特色旅游、商业品牌项目，拟打造为"文化 + 旅游 + 产业"的现代特色产业集聚区。

Tingyin Lake District is the center of building tourism highland of "Extraordinary Qiaoshan with Distinctive Lingnan Culture". This district as an extension area will center on the introduction of cultural creativity, special tourism and commercial brands, to create a modern of industrial clusters of with a feature of Culture + Tourism + Industries.

### 山南片区：互联网 + 科创产业生态社区
### South Mountain District : internet + scientific and technological innovation industries

片区定位是以产业互联网为核心，以智能应用为主导，重点引入智能制造、工业互联网、产业大数据、设计创新等高新技术企业，打造以"高技术制造、研发中试、检验检测、展览展示"为核心功能的科技创新产业载体平台。

With industrial internet as the core and intelligent applications as a leading role, this district puts priority on the introduction of high-tech enterprises in fields of intelligent manufacturing, industrial internet, industrial data and design innovation, etc., to create a platform for industrial S&T innovation, within which "high-tech manufacturing, trials in research and development, inspection and detection, exhibition and presentation" serve as the core functions.

### 三乡片区：新消费品 + 泛家居产业社区
### Sanxiang District : new consumer goods + comprehensive home furnishing industry

片区产业定位以机械制造、卫生用品、新消费品、电商服务平台四大产业为发展方向，通过三旧改造，整合存量载体资源，促进土地利用集约化，拓展智能制造空间。

With four industries of machinery manufacturing, sanitary products, new consumer goods and e-commerce service platform as the development direction, this district integrates the stock carrier resources through "Three Old Transformation" programme, promotes the intensification of land use and expands the space for intelligent manufacturing.

### 西江片区：工业互联网产业社区 +IT 数据产业园
### Xijiang District : industrial internet industry community + IT data industrial park

片区未来产业定位是以计算机、通信设备制造、通信服务和大数据服务为方向，打造集数据流、技术流、业务流、人才流、资金流于一体的新时代数字经济产业社区，助推产业转型升级。

This district will be led by computer and communication equipment manufacturing, communication services and big data services, to build a new digital economy industrial community integrating data flow, technology flow, business flow, talent flow and capital flow, and to boost industrial transformation and upgrading.

# 城乡融合 · 宜居西樵

## Livable Xiqiao with Urban-Rural Integration

聚焦广东省城乡融合发展改革创新实验区建设，全面实施乡村振兴战略，重点打造环西樵山片区，实现生产、生活、生态的融合发展，进一步提升西樵镇城乡融合发展水平。

Xiqiao focuses on the construction of reform and innovation pilot zone for urban-rural integration of Guangdong Province, fully implement the rural revitalization strategy and mainly construct the area around Xiqiao Mountain, to achieve the integrated development of production, life and ecology and further enhance the level of urban-rural integration development in Xiqiao.

打造城市更新优秀典范（官山人家）；打造乡村振兴连片示范区（特色精品示范村儒溪村）；建设百里芳华示范片区（渔耕粤韵文化旅游园）；推进村级工业园连片改造（五八科创园）。

Setting up a Prime Example of Urban Renewal (Artist's impression of Guanshan Residence).Creating a Contiguous Demonstration Zone of Rural Revitalization High-quality demonstration village with distinctive features (Ruxi Village). Building Demonstration Zone of Baili Fanghua (Cantonese fishery and agro-farming culture tourist park). Promoting Contiguous Renovation of Village-level Industrial Parks (Five-eighth science and technology innovation park ).

# 绿色发展 · 美丽西樵
## Beautiful Xiqiao with Green Development

践行绿水青山就是金山银山的发展理念，以创建国家生态公园和建设高品质森林小镇为抓手，加强生态保护和修复，努力打造"青山常在、绿水长流、空气常新"的美丽西樵，为大湾区高品质森林城市建设贡献西樵力量。

In order to pursue the development concept of Clear waters and green mountains are as good as mountains of gold and silver, and with the construction of national ecological park and high-quality forest town as the starting point, Xiqiao will strengthen ecological protection and restoration and strive to create a beautiful Xiqiao with green mountains, clean waters and fresh air and make contribution to the construction of high-quality forest city in the Greater Bay Area.

**中国重要农业文化遗产 —— 广东佛山珠三角基塘农业系统**
**China's Important Agricultural Heritage–Dike–pond Agricultural System in the Pearl River Delta (Foshan, Guangdong)**

利用西樵山下万亩桑基鱼塘及山、水、林、田、湖、岛生态资源，建设"生态、生活、生产"和文旅相融合的生态文明示范区，打造大湾区绿色明珠。

With thousands of acres of mulberry-based ponds at the foot of Xiqiao Mountain and rich ecological resources of mountains, water, forests, fields, lakes and islands, Xiqiao will build an ecological civilization demonstration zone for cultural tourism, integrating ecology, life and production and creating a green pearl in the Greater Bay Area.